National 5

MATHS

PRACTICE QUESTION BOOK

N5 MATHS PRACTICE QUESTION BOOK

Craig Lowther • Judith Walker

ISBN 9780008209087

Published by
Leckie & Leckie Ltd
An imprint of HarperCollinsPublishers
Westerhill Road, Bishopbriggs, Glasgow, G64 2QT
T: 0844 576 8126   F: 0844 576 8131
leckieandleckie@harpercollins.co.uk   www.leckieandleckie.co.uk

Commissioning Editor: Clare Souza
Managing Editor: Craig Balfour

Special thanks to
Jouve (layout); Ink Tank (cover design);
Maggie Donovan (proofread);
Project One Publishing Solutions, Scotland (project management and editing)

This book is based on original material written by Kath Hipkiss (Collins AQA GCSE Maths Foundation Practice Book) and Rob Ellis (Collins AQA GCSE Maths Higher Practice Book).

A CIP Catalogue record for this book is available from the British Library.

Acknowledgements
Whilst every effort has been made to trace the copyright holders, in cases where this has been unsuccessful, or if any have inadvertently been overlooked, the Publishers would gladly receive any information enabling them to rectify any error or omission at the first opportunity.

Printed in Italy by Grafica Veneta S.P.A.

MIX
Paper from
responsible sources
FSC™ C007454

FSC™ is a non-profit international organisation established to promote the responsible management of the world's forests. Products carrying the FSC label are independently certified to assure consumers that they come from forests that are managed to meet the social, economic and ecological needs of present and future generations, and other controlled sources.

Find out more about HarperCollins and the environment at
www.harpercollins.co.uk/green

How to use this book

ANSWERS
www.leckieandleckie.co.uk/page/Resources

How to use this book

Welcome to Leckie and Leckie's *National 5 Maths Practice Book*. This book follows the structure of the Leckie and Leckie *National 5 Maths Student Book*, so is ideal to use alongside it. Questions have been written to provide practice for topics and concepts which have been identified as challenging for many students.

Examples

Examples with worked solutions provide support for particularly tricky concepts.

Use of calculators

Questions when you should **not** use a calculator are marked with a icon.

Questions when you could use a calculator are marked with a icon.

Reasoning questions

Questions which require reasoning skills are marked with a icon.

Hints

Where appropriate, hints are provided to help give extra guidance and support.

Answers

Check your own work. The answers are provided online at:

www.leckieandleckie.co.uk/page/Resources

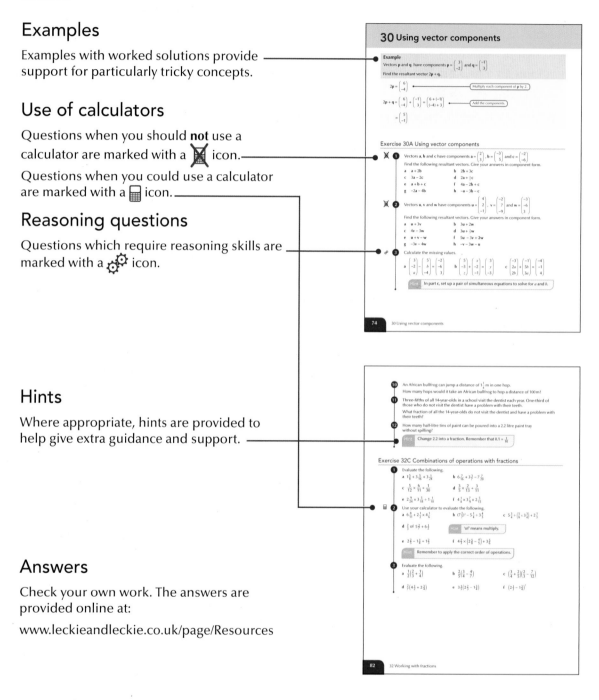

1 Working with surds

Exercise 1A Surds

1 Simplify, giving your answer in surd form where necessary.

a $\sqrt{3} \times \sqrt{4}$ **b** $\sqrt{5} \times \sqrt{7}$ **c** $\sqrt{5} \times \sqrt{5}$ **d** $\sqrt{2} \times \sqrt{32}$

2 Simplify, giving your answer in surd form where necessary.

a $\sqrt{15} \div \sqrt{5}$ **b** $\sqrt{18} \div \sqrt{2}$ **c** $\sqrt{32} \div \sqrt{2}$ **d** $\sqrt{12} \div \sqrt{8}$

3 Simplify, giving your answer in surd form where necessary.

a $\sqrt{90}$ **b** $\sqrt{32}$ **c** $\sqrt{63}$ **d** $\sqrt{300}$

e $\sqrt{150}$ **f** $\sqrt{270}$ **g** $\sqrt{96}$ **h** $\sqrt{125}$

4 Simplify each expression.

a $2\sqrt{5} \times 5\sqrt{3}$ **b** $2\sqrt{3} \times 3\sqrt{3}$ **c** $2\sqrt{2} \times 3\sqrt{8}$ **d** $2\sqrt{3} \times 2\sqrt{27}$

e $5\sqrt{18} \div \sqrt{2}$ **f** $2\sqrt{32} \div 4\sqrt{8}$ **g** $5\sqrt{2} \times \sqrt{8} \div 2\sqrt{2}$ **h** $\sqrt{32} \times \sqrt{8}$

5 Simplify these expressions.

a $\sqrt{32} + \sqrt{8}$ **b** $\sqrt{50} + \sqrt{8}$ **c** $\sqrt{27} - \sqrt{12}$

d $\sqrt{8} + \sqrt{72} - \sqrt{2}$ **e** $\sqrt{80} - \sqrt{45} + \sqrt{20}$ **f** $\sqrt{18} + \sqrt{12} + \sqrt{32} + \sqrt{75}$

g $\sqrt{24} + \sqrt{20} + \sqrt{54} - \sqrt{125}$ **h** $\sqrt{50} - \sqrt{48} + \sqrt{18} - \sqrt{12}$

6 Expand and simplify where possible.

a $\sqrt{5}(3 - \sqrt{2})$ **b** $\sqrt{8}(3 - 4\sqrt{2})$ **c** $3\sqrt{8}(2\sqrt{2} + 4)$

d $(2 + \sqrt{3})(1 - \sqrt{3})$ **e** $(3 + \sqrt{5})(2 - \sqrt{5})$ **f** $(3 - \sqrt{2})(4 + 2\sqrt{2})$

7 Rationalise the denominators of each expression.

a $\dfrac{1}{\sqrt{7}}$ **b** $\dfrac{2}{\sqrt{5}}$ **c** $\dfrac{2}{\sqrt{8}}$ **d** $\dfrac{1}{2\sqrt{2}}$

e $\dfrac{5\sqrt{3}}{\sqrt{27}}$ **f** $\dfrac{\sqrt{8}}{\sqrt{3}}$ **g** $\dfrac{1 + \sqrt{3}}{\sqrt{3}}$ **h** $\dfrac{3 - \sqrt{2}}{\sqrt{8}}$

8 The function f(x) is described by f(x) = $\dfrac{4}{\sqrt{x}}$, $x > 0$.

Find the value of the following, expressing your answer with a rational denominator.

a f(7)   **b** f(6)   **c** f(8)

9 Find the length of the side marked x in these triangles. Give your answer as a surd in its simplest form.

a

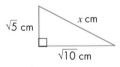

b

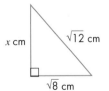

10 Calculate the area of these rectangles. Give your answer as a surd in its simplest form.

a

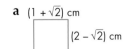

b

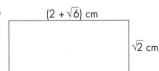

2 Simplifying expressions using the laws of indices

Exercise 2A Indices

1 Simplify each expression.

a $x^3 \times x^7$ **b** $x^5 \times x^4$ **c** $x^5 \times x^2$ **d** $x^3 \times x^2$

e $x^6 \times x^1$ **f** $x^7 \times x^5$ **g** $x^7 \times x^4$ **h** $x^4 \times x^9$

i $x^3 \times x^7 \times x^1$ **j** $x^5 \times x^3 \times x^8$ **k** $x^4 \times x \times x^3$ **l** $x^4 \times x^{12} \times x^5 \times x^2$

2 Simplify each expression.

a $y^{10} \div y^3$ **b** $y^4 \div y^3$ **c** $y^7 \div y$ **d** $y^2 \div y^2$

e $y^{20} \div y^4$ **f** $y^3 \div y$ **g** $y^4 \div y^2$ **h** $y^{15} \div y^3$

i $\dfrac{y^8 \times y^2}{y^7}$ **j** $\dfrac{y^9 \times y^4}{y^6}$ **k** $\dfrac{y^5 \times y^{21} \times y^7}{y^9}$ **l** $\dfrac{y^3 \times y^6 \times y^9}{y^3 \times y^5}$

3 Simplify each expression.

a $3a^4 \times 5a^2$ **b** $3a^4 \times 7a$ **c** $5a^4 \times 6a^2$ **d** $3a^2 \times 4a^7$

e $5a^4 \times \left(-5a^2\right) \times 5a^2$ **f** $2a^3 \times 3a^4 \times 6a^5$

g $3a^4 \times \left(-6a^2\right) \times 5a^7$ **h** $4a^5 \times \left(-2a^3\right) \times \left(-7a^6\right)$

4 Simplify each expression.

a $30a^6 \div 5a^5$ **b** $15a^5 \div 3a^5$ **c** $15a^5 \div 5a$ **d** $18a^5 \div 3a^1$

e $30a \div 3a^5$ **f** $\dfrac{72a^5}{9a^3}$ **g** $\dfrac{40a^7}{8a^3}$ **h** $\dfrac{48a^4}{4a^6}$

5 Simplify each expression.

a $5a^5b^3 \times 7a^3b$ **b** $3a^5b^3 \times 5a^{-7}b^{-5}$ **c** $15a^5b^7 \div 3ab$ **d** $57a^{-3}b^7 \div 3a^5b^{-3}$

e $14a^7b^{-5} \div 7a^5b^3$ **f** $\dfrac{32a^2b^{-3}}{8ab^2}$ **g** $\dfrac{45a^2b^3}{9a^{-4}b^6}$ **h** $\dfrac{50a^3b^5}{15a^4b^{-2}}$

6 Simplify each expression.

a $\left(x^3\right)^4$ **b** $\left(x^4\right)^{-5}$ **c** $\left(2x^5\right)^3$ **d** $\left(3x^{-2}\right)^4$

e $\left(x^5\right)^3 \times x^4$ **f** $\left(x^{-4}\right)^2 \times \left(x^3\right)^4$ **g** $\left(3x^3\right)^2 \times \left(2x^{-1}\right)^4$ **h** $\left(5x^{-3}\right)^2 \times \left(2x^{-4}\right)^3$

7 Simplify each expression.

a $\dfrac{4x^6 \times 3x^{-2}}{2x^2}$ **b** $\dfrac{5x^3 \times 8x^4}{10x^{-3}}$ **c** $\dfrac{7x^5 \times 3x^3}{6x^8}$ **d** $\dfrac{5x^{-4} \times \left(2x^4\right)^3}{10x^2}$

8 Write down each number with a positive power.

a 5^{-2} **b** 4^{-1} **c** 10^{-3} **d** 3^{-3}

9 Simplify, giving your answer with a positive power.

a x^{-2} **b** y^{-5} **c** $5t^{-1}$

d $6n^{-4}$ **e** $\dfrac{f^{-3}}{2}$ **f** $\dfrac{3h^{-5}}{4}$

10 Find the value of each expression for the value of the letter shown.

a $x = 3$ i x^2 ii $4x^{-1}$

b $t = 5$ i t^{-2} ii $5t^{-4}$

c $m = 2$ i m^{-3} ii $4m^{-2}$

11 Calculate the value of these expressions when $a = 3$ and $b = 2$. Give each answer as a fraction in its simplest form.

a $3a^{-1} + 2b^{-2}$ b $a^{-2} + b^{-3}$

12 Rewrite each number in index form.

a $\sqrt[4]{t^3}$ b $\sqrt[5]{m^2}$

13 Evaluate each number.

a $36^{\frac{1}{2}}$ b $144^{\frac{1}{2}}$ c $25^{\frac{1}{2}}$ d $196^{\frac{1}{2}}$ e $8^{\frac{1}{3}}$

f $125^{\frac{1}{3}}$ g $32^{\frac{1}{5}}$ h $81^{\frac{1}{4}}$ i $27^{\frac{1}{3}}$ j 6^0

14 Evaluate each number.

a $16^{\frac{3}{4}}$ b $125^{\frac{4}{3}}$ c $81^{\frac{3}{4}}$

d $27^{\frac{2}{3}}$ e $8^{\frac{4}{3}}$ f $36^{\frac{3}{2}}$

Exercise 2B Scientific notation

1 Write these numbers out in full.

a 2.5×10^2 b 3.12×10 c 4.32×10^{-3} d 2.43×10

e 2.0719×10^{-2} f 5.372×10^3 g 2.03×10^2 h 1.3×10^3

i 8.17×10^5 j 8.35×10^{-3} k 3×10^7 l 5.27×10^{-4}

2 Write these numbers in scientific notation.

a 200 b 0.305 c $40\ 700$ d $3\ 400\ 000\ 000$

e $20\ 780\ 000\ 000$ f $0.000\ 537\ 8$ g 2437 h 0.173

3 Write the number in each statement in scientific notation.

a Last year there were 24 673 000 vehicles licensed in the UK.

b Daryl John was one of 15 282 runners to complete the Boston Marathon.

c Last year 613 000 000 000 passenger kilometres were completed on British roads.

d The Sun is 93 million miles away from Earth. The next closest star to the Earth is Proxima Centauri at a distance of about 24 million million miles.

e A scientist claims to be working with a new particle weighing only 0.000 000 000 000 65 g.

4 Calculate, and express your answers in scientific notation.

a $3.1 \times 10^2 \times 5.1 \times 10^3$ b $2.4 \times 10^5 \times 3.2 \times 10^3$ c $9.6 \times 10^5 \times 7.6 \times 10^3$

d $6.3 \times 10^{-4} \times 3.4 \times 10^2$ e $(2.1 \times 10^5)^2$ f $(7.8 \times 10^{-3})^2$

5 Calculate, and express your answers in scientific notation.

   a $(9 \times 10^8) \div (3 \times 10^4)$ **b** $(2.7 \times 10^7) \div (9 \times 10^3)$

   c $(5.5 \times 10^4) \div (1.1 \times 10^{-2})$ **d** $(4.2 \times 10^{-9}) \div (3 \times 10^{-8})$

6 Work out, and express your answers in scientific notation.

   a $\dfrac{8 \times 10^9}{4 \times 10^7}$ **b** $\dfrac{12 \times 10^6}{3 \times 10^4}$ **c** $\dfrac{2.8 \times 10^7}{7 \times 10^{-4}}$

7 The population of Africa in 2013 was approximately 1×10^9.

It is expected to reach 1.8×10^9 in 2050.

By how much is the population of Africa expected to rise between 2013 and 2050?

Give your answer in scientific notation.

8 The speed of light is approximately 3.00×10^8 m/s.

It takes about 1.3 seconds for light to travel to the Moon.

Work out the approximate distance to the Moon in kilometres.

Give your answer in scientific notation.

9 The Moon is a sphere with a radius of 1.08×10^3 miles.

Calculate the volume of the Moon.

Give your answer in scientific notation, correct to 3 significant figures.

> **Hint** The formula for the volume, V, of a sphere is:
>
> $V = \frac{4}{3}\pi r^3$

10 How many seconds are there in a leap year? Give your answer in scientific notation.

11 The starship *Enterprise* travelled 3×10^{15} km at a speed of 1.5×10^3 km/s.

How long did the journey take?

Give your answer to the nearest year.

12 The average distance from Earth to Mars is 2.25×10^8 km.
A video message is sent to Mars at a speed of 1.8×10^7 km/min.

How long will it take to arrive?

13 The average ant weighs 3×10^{-4} g.
What is the approximate weight of 150 ants?

Give your answer in scientific notation.

14 The volume of Venus is 9.3×10^{11} km³. The volume of Pluto is 7.2×10^9 km³.

Calculate how many times bigger the volume of Venus is than the volume of Pluto.

Give your answer to the nearest whole number.

15 $x = 1.8 \times 10^7$ and $y = 4 \times 10^3$.

Express $y - 3x$ in scientific notation.

3 Working with algebraic expressions involving expansion of brackets

Exercise 3A Expanding brackets

1 Expand each expression.

a $3(m + 7)$ **b** $2(x - y)$ **c** $x(x + 6)$

d $a(a - b)$ **e** $4x(x - 7)$ **f** $2y(3y + 2k)$

2 Expand and simplify each expression.

a $3(2x - 1) + 2(5x + 3)$ **b** $4(3y - 2) - 3(2y - 5)$ **c** $2x(3x - 5) - 2(4x + 1)$

d $8 + 3(2x - 7)$ **e** $5x(3x - 4) - 8x$ **f** $8 - 2(2x + 3) - x(3x - 5)$

3 Expand and simplify.

a $(x + 2)(x + 4)$ **b** $(x - 3)(x + 1)$ **c** $(x + 4)(x - 1)$ **d** $(x - 5)(x - 2)$

e $(x + 3)(x - 3)$ **f** $(x - 3)(x - 3)$ **g** $(x + 6)(x + 1)$ **h** $(x - 6)(x - 1)$

4 For the following:

i find the mistake(s) made in each expansion

ii write the correct answer.

a $(x + 3)(x + 2) = x^2 + 5x + 5$ **b** $(x + 11)(x - 7) = x^2 + 18x + 77$

c $(x - 2)(x + 9) = x^2 - 7x + 18$ **d** $(x - 2)(x - 12) = x^2 - 10x + 24$

5 Expand and simplify.

a $(3x + 4)(4x + 2)$ **b** $(2y + 1)(3y + 2)$ **c** $(4t + 2)(3t + 6)$

d $(3t + 2)(2t - 1)$ **e** $(6m + 1)(3m - 2)$ **f** $(5k + 3)(4k - 3)$

g $(4p - 5)(3p + 4)$ **h** $(6w + 1)(3w + 4)$ **i** $(3a - 4)(5a + 1)$

j $(5r - 2)(3r - 1)$ **k** $(4g - 1)(3g - 2)$ **l** $(3d - 2)(4d + 1)$

m $(3 + 4p)(5 + 4p)$ **n** $(3 + 2t)(5 + 3t)$ **o** $(2 + 5p)(3p + 1)$

6 Expand and simplify.

a $(x + 1)^2$ **b** $(x - 2)^2$ **c** $(x - 9)^2$

d $(x + 3)^2$ **e** $(2x - 9)^2$ **f** $(a + b)^2$

g $(a - b)^2$ **h** $(m - 2n)^2$ **i** $(x + y)^2$ **j** $(2a + 3b)^2$ **k** $(3a - 6b)^2$

> **Hint** $(x + 1)^2 = (x + 1)(x + 1)$

7 Expand and simplify.

a $(x + 2)(x + 3)(x + 4)$ **b** $(x + 1)(x - 3)(x + 2)$ **c** $(x - 3)(x + 3)(x - 5)$

d $(x - 4)^2(x + 3)$ **e** $(x^2 + 2x + 1)(x - 3)$ **f** $(x^2 - 3x - 2)(x - 7)$

8 Expand and simplify.

a $(x + 3)(x^2 - 2x + 5)$ **b** $(2x + 1)(x^2 + 2x - 3)$

c $(3x - 2)(2x^2 - 3x + 5)$ **d** $(4x - 1)(3x^2 - 2x - 7)$

e $(2x - 3)(x + 4) + (3x - 2)(x - 3)$ **f** $(4x - 5)(2x - 3) - (2x - 1)(x + 3)$

g $(5x + 2)(3x - 4) - (3x + 2)^2$ **h** $(2x - 5)^2 - (3x + 2)^2$

i $(3x - 2)(2x^2 - 3x + 5) - (2x - 3)(x + 5)$

4 Factorising an algebraic expression

Exercise 4A Factorising

1 Factorise each expression.

  a $8ab - 6bc$ **b** $4a^2 - 8ab$ **c** $8mt - 6pt$

  d $20at^2 + 12at$ **e** $4b^2c - 10bc$ **f** $4abc + 6bcd$

  g $6a^2 + 4a + 10$ **h** $12ab + 6bc + 9bd$ **i** $6t^2 + 3t + at$

  j $96mt^2 - 3mt + 69m^2t$ **k** $6ab^2 + 2ab - 4a^2b$ **l** $5pt^2 + 15pt + 5p^2t$

2 Factorise each expression.

  a $x^2 + 7x + 6$ **b** $t^2 + 4t + 4$ **c** $m^2 + 11m + 10$ **d** $k^2 + 11k + 24$

  e $k^2 - 10k + 21$ **f** $f^2 - 22f + 21$ **g** $m^2 - 5m + 4$ **h** $p^2 - 7p + 10$

  i $y^2 + y - 6$ **j** $t^2 + 7t - 8$ **k** $x^2 + 9x - 10$ **l** $r^2 + 6r - 7$

  m $n^2 - 7n - 18$ **n** $m^2 - 20m - 44$ **o** $x^2 - x - 72$ **p** $t^2 - 18t - 63$

3 Factorise each expression.

  a $3x^2 + 4x + 1$ **b** $3x^2 - 2x - 1$ **c** $4x^2 + 8x + 3$ **d** $2x^2 + 7x + 3$

  e $15x^2 + 13x + 2$ **f** $4x^2 + 4x - 3$ **g** $6x^2 - 7x + 2$ **h** $8x^2 - 8x - 6$

  i $8x^2 - 13x - 6$ **j** $6x^2 - 13x + 2$ **k** $10x^2 + 11x - 6$ **l** $6x^2 + 11x - 2$

4 Factorise each expression.

  a $x^2 - 81$ **b** $t^2 - 36$ **c** $4 - x^2$ **d** $81 - t^2$

  e $k^2 - 400$ **f** $64 - y^2$ **g** $x^2 - y^2$ **h** $a^2 - 9b^2$

  i $9x^2 - 25y^2$ **j** $9x^2 - 16$ **k** $100t^2 - 4w^2$ **l** $36a^2 - 49b^2$

5 Three students are asked to factorise the expression $4x^2 + 4x - 8$.

  These are their answers:

  Adriana   Ben   Cara

  $(2x + 4)(2x - 2)$ $(4x + 8)(x - 1)$ $(x + 2)(4x - 4)$

  All the answers are correctly factorised.

  a Show that one quadratic expression can have three different factorisations.

  b Which of the following is the most complete factorisation? Give reasons for your choice.

    $2(x + 2)(2x - 2)$ $4(x + 2)(x - 1)$ $2(2x + 4)(x - 1)$

6 Fully factorise.

  a $3x^2 + 3x - 18$ **b** $5x^2 - 20$ **c** $4x^2 - 8xy$

  d $4x^2 + 6x - 4$ **e** $27x^2 - 12$ **f** $2x^3 - 14x^2 + 24x$

  g $3x^3 - 48x$ **h** $6x^2y - 13xy + 6y$

5 Completing the square in a quadratic expression with unitary x^2 coefficient

Example

Express $x^2 - 6x + 2$ in the form $(x + a)^2 + b$.

$x^2 - 6x + 2 = (x - 3)^2 - 9 + 2$

$\qquad\qquad = (x - 3)^2 - 7$

> Divide the coefficient of x, 6, by 2. Multiply out $(x - 3)^2$ to give $x^2 - 6x + 9$ and subtract 9, in order to keep the value of the expression correct: $x^2 - 6x + 9 - 9$.

Exercise 5A Completing the square

1 Express each of the following in the form $(x + p)^2 + q$.

  a $x^2 + 16x$ **b** $x^2 - 8x$ **c** $x^2 + 6x$ **d** $x^2 - 12x$

  e $x^2 - 20x$ **f** $x^2 + 18x$ **g** $x^2 - 30x$ **h** $x^2 + 5x$

2 Each expression below can be written in the form $(x + a)^2 + b$.

For each expression, find the values of a and b.

  a $x^2 - 2x$ **b** $x^2 + 10x$ **c** $x^2 - 14x$ **d** $x^2 + 12x$

  e $x^2 - 22x$ **f** $x^2 + 80x$ **g** $x^2 - x$ **h** $x^2 + 11x$

3 Express each of the following in the form $(x + p)^2 + q$.

  a $x^2 + 4x + 9$ **b** $x^2 + 8x + 20$ **c** $x^2 - 2x + 7$ **d** $x^2 - 12x + 31$

  e $x^2 + 6x - 2$ **f** $x^2 - 10x - 4$ **g** $x^2 - 16x + 59$ **h** $x^2 + x + 1$

  i $x^2 + 4x - 2$ **j** $x^2 + 10x - 3$ **k** $x^2 - 6x + 15$ **l** $x^2 - 12x - 3$

  m $x^2 + 2x - 5$ **n** $x^2 - 18x + 57$ **o** $x^2 - 5x + 2$ **p** $x^2 - 4x - 10$

  q $x^2 + 12x + 40$ **r** $x^2 - 10x + 13$ **s** $x^2 + 6x + 11$ **t** $x^2 - 2x - 9$

4 Each expression below can be written in the form $(x + a)^2 + b$.

For each expression, find the values of a and b.

  a $x^2 - 4x + 7$ **b** $x^2 + 10x - 2$ **c** $x^2 - 6x - 3$ **d** $x^2 - 14x + 37$

  e $x^2 + 20x - 43$ **f** $x^2 - 18x + 93$ **g** $x^2 - 60x + 185$ **h** $x^2 + 3x + 5$

  i $x^2 - 6x - 4$ **j** $x^2 + 2x + 9$ **k** $x^2 - 14x + 50$ **l** $x^2 - 22x + 130$

  m $x^2 - 20x + 29$ **n** $x^2 - 10x + 18$ **o** $x^2 + 7x + 12$ **p** $x^2 + 40x + 500$

  q $x^2 - 8x - 17$ **r** $x^2 + 6x - 20$ **s** $x^2 - 2x - 14$ **t** $x^2 - x - 3$

Example

Solve this equation by completing the square.

$x^2 + 6x + 2 = 0$

Round your answers to 1 decimal place (1 d.p.).

$(x + 3)^2 - 9 + 2 = 0$ Complete the square on the left-hand side.

$(x + 3)^2 - 7 = 0$

$(x + 3)^2 = 7$ Rearrange the equation.

$x + 3 = \pm\sqrt{7}$

$x + 3 = \sqrt{7}$ or $x + 3 = -\sqrt{7}$

$x = \sqrt{7} - 3$ or $x = -\sqrt{7} - 3$ Leave your answer like this if you are asked for the answer in surd form.

$x = -0.4$ or $x = -5.6$ (1 d.p.)

5 Solve each equation by completing the square. Leave your answers in surd form.

a $x^2 - 2x - 2 = 0$ b $x^2 + 6x - 1 = 0$ c $x^2 - 10x + 18 = 0$

d $x^2 - 4x + 2 = 0$ e $x^2 + 10x - 4 = 0$ f $x^2 + 8x + 13 = 0$

g $x^2 + 6x - 4 = 0$ h $x^2 - 2x - 2 = 0$ i $x^2 + 12x + 30 = 0$

6 Solve each equation by completing the square. Round the answers to 1 d.p.

a $x^2 - 4x + 1 = 0$ b $x^2 + 12x + 28 = 0$ c $x^2 - 8x + 13 = 0$

d $x^2 - 10x + 23 = 0$ e $x^2 + 6x - 5 = 0$ f $x^2 - 5x + 3 = 0$

6 Reducing an algebraic fraction to its simplest form

Exercise 6A Simplest form

1 Simplify each of the following expressions.

a $\dfrac{6a}{9}$

b $\dfrac{8m}{14m}$, $m \neq 0$

c $\dfrac{6c}{18d}$, $d \neq 0$

d $\dfrac{9g^2h}{12gk}$, $g \neq 0$, $k \neq 0$

e $\dfrac{21p^2q}{14pq^2}$, $p \neq 0$, $q \neq 0$

f $\dfrac{18pqr}{22prt}$, $p \neq 0$, $r \neq 0$, $t \neq 0$

g $\dfrac{20a^2bc}{16abc^2}$, $a \neq 0$, $b \neq 0$, $c \neq 0$

h $\dfrac{48gh^2m}{16gh}$, $g \neq 0$, $h \neq 0$

> **Hint** If the denominator is cancelled completely, do **not** leave the answer with a 1 in the denominator.

2 Simplify each of the following expressions.

a $\dfrac{x(x+1)}{5(x+1)}$, $x \neq -1$

b $\dfrac{(x+2)(x-1)}{(x-1)(2x+5)}$, $x \neq 1$, $x \neq -\dfrac{5}{2}$

c $\dfrac{(2x-3)(x+2)}{(x+2)^2}$, $x \neq -2$

d $\dfrac{x(x+3)}{4(x+3)}$, $x \neq -3$

e $\dfrac{(3x-2)(3x+1)}{4(3x+1)(2x+1)}$, $x \neq -\dfrac{1}{3}$, $x \neq -\dfrac{1}{2}$

f $\dfrac{2(5x-2)^2}{4(5x-2)(x+3)}$, $x \neq \dfrac{2}{5}$, $x \neq -3$

> **Hint** Make sure you don't try to simplify further than you can. For example, $\dfrac{x+2}{x+1}$ does not cancel further to $\dfrac{2}{1}$.

Example

Simplify $\dfrac{x^2-16}{2x^2-5x-12}$, $x \neq -\dfrac{3}{2}$, $x \neq 4$

$x^2 - 16 = (x-4)(x+4)$ — Factorise the numerator using the difference of two squares.

$2x^2 - 5x - 12 = (2x+3)(x-4)$ — Factorise the trinomial in the denominator.

$\dfrac{x^2-16}{2x^2-5x-12} = \dfrac{(x-4)(x+4)}{(2x+3)(x-4)}$ — Simplify by cancelling out brackets or factors common to both numerator and denominator.

$= \dfrac{x+4}{2x+3}$ — Write down the remaining terms.

3 Simplify each of the following expressions.

a $\dfrac{5x + 10}{x^2 + 2x}$, $x \neq 0$, $x \neq -2$

b $\dfrac{2x^2 - 6x}{8x^2 - 2x}$, $x \neq 0$, $x \neq \dfrac{1}{4}$

> **Hint** Factorise the numerator and denominator, then simplify.

c $\dfrac{x^2 - 4}{4x^2 - 8x}$, $x \neq 0$, $x \neq 2$

d $\dfrac{x^2 - x - 6}{x^2 + 2x - 15}$, $x \neq 3$, $x \neq -5$

e $\dfrac{2x^2 + x - 6}{x^2 - 2x - 8}$, $x \neq 4$, $x \neq -2$

f $\dfrac{4x^2 - 25}{6x^2 + 13x - 5}$, $x \neq -\dfrac{5}{2}$, $x \neq \dfrac{1}{3}$

g $\dfrac{2x^2 - 18}{3x^2 + 6x - 9}$, $x \neq 1$, $x \neq -3$

h $\dfrac{x^3 + 3x^2 - 10x}{x^3 - 25x}$, $x \neq 0$, $x \neq -5$, $x \neq 5$

i $\dfrac{(x + 4)^3}{(x^2 - 16)(x + 4)}$, $x \neq -4$, $x \neq 4$

j $\dfrac{2x^2 - x - 3}{4x^2 + 4x - 15}$, $x \neq -\dfrac{5}{2}$, $x \neq \dfrac{3}{2}$

k $\dfrac{x^2 - 4}{2 - x}$, $x \neq 2$

l $\dfrac{x^2 - 36}{12 + 4x - x^2}$, $x \neq -2$, $x \neq 6$

4
a Factorise $x^2 + x - 12$

b Hence simplify $\dfrac{x^2 + x - 12}{x^2 - 9}$, $x \neq 3$, $x \neq -3$

5
a Factorise $4x^2 - 25$

b Hence simplify $\dfrac{4x^2 - 25}{10x^2 + 25x}$, $x \neq 0$, $x \neq -\dfrac{5}{2}$

6
a Factorise $3x^2 + 11x - 4$

b Hence simplify $\dfrac{3x^2 + 11x - 4}{2x^2 - 32}$, $x \neq 4$, $x \neq -4$

7
a Factorise $4x^2 + 7x - 2$

b Hence simplify $\dfrac{4x^2 + 7x - 2}{3x^2 + 5x - 2}$, $x \neq -2$, $x \neq \dfrac{1}{3}$

7 Applying one of the four operations to algebraic fractions

Exercise 7A Operations on algebraic fractions

1 Express each of the following as a fraction in its simplest form.

a $\dfrac{5}{x} \times \dfrac{x}{15}$, $x \neq 0$

b $\dfrac{3x}{7y} \times \dfrac{14y^2}{15}$, $y \neq 0$

c $\dfrac{8a^2}{9by} \times \dfrac{3b}{2a}$, $a \neq 0$, $b \neq 0$, $y \neq 0$

d $\dfrac{3x}{2} \times \dfrac{4x}{5}$

e $\dfrac{x+1}{4} \times \dfrac{3}{2x+2}$

f $\dfrac{2x-1}{2} \times \dfrac{4}{3x-1}$

g $\dfrac{3x+6}{x^2-3x+2} \times \dfrac{x^2-4}{x^2+2x}$, $x \neq 1$, $x \neq -1$, $x \neq 2$

h $\dfrac{x^2-5x+6}{x^2+2x} \times \dfrac{2x^2+3x-2}{x^2-9}$ $x \neq -2$, $x \neq -3$, $x \neq 3$

Example

Express $\dfrac{7h}{m} \div \dfrac{h}{4m^2}$ as a single fraction in its simplest form.

$\dfrac{7h}{m} \div \dfrac{h}{4m^2} = \dfrac{7h}{m} \times \dfrac{4m^2}{h}$

> Multiply by the reciprocal of the second fraction. 'Reciprocal' means swap the numerator and denominator.

$= \dfrac{7\cancel{h}^{\,1}}{\cancel{m}_{\,1}} \times \dfrac{4m^{2\,\,m}}{\cancel{h}_{\,1}}$

> Cancel the common factors in the numerator and denominator.

$= \dfrac{7}{1} \times \dfrac{4m}{1}$

$= \dfrac{28m}{1}$

> Make sure that you reduce the fraction to its simplest form.

$= 28m$

2 Express each of the following as a fraction in its simplest form.

a $\dfrac{6}{y} \div \dfrac{9}{y}$, $y \neq 0$

b $\dfrac{2}{g} \div \dfrac{g}{t}$, $g \neq 0$, $t \neq 0$

c $\dfrac{5xy}{6} \div \dfrac{3y}{2x}$, $x \neq 0$, $y \neq 0$

d $\dfrac{x}{4} \div \dfrac{2x}{5}$

e $\dfrac{x+3}{2} \div \dfrac{2x+6}{5}$

f $\dfrac{4x-2}{3} \div \dfrac{2x-1}{4}$

g $\dfrac{2x+2}{x^2-9} \div \dfrac{x^2+x}{x^2+2x-3}$, $x \neq 1$, $x \neq -1$ $x \neq -3$, $x \neq 3$

h $\dfrac{xy+2y}{x^2-7x+12} \div \dfrac{5x+10}{x^2-16}$, $x \neq 4$, $x \neq -4$, $x \neq 3$

Example

Express as a fraction in its simplest form: $\dfrac{4}{x+2} - \dfrac{3}{x-1}$

$$\dfrac{4}{x+2} - \dfrac{3}{x-1} = \dfrac{4}{(x+2)} \times \dfrac{(x-1)}{(x-1)} - \dfrac{3}{(x-1)} \times \dfrac{(x+2)}{(x+2)}$$

> Multiply the first fraction by $\dfrac{x-1}{x-1}$ and the second fraction by $\dfrac{x+2}{x+2}$ to give a common denominator of $(x+2)(x-1)$.

$$= \dfrac{4(x-1)}{(x+2)(x-1)} - \dfrac{3(x+2)}{(x+2)(x-1)}$$

$$= \dfrac{4(x-1) - 3(x+2)}{(x+2)(x-1)}$$

$$= \dfrac{4x - 4 - 3x - 6}{(x+2)(x-1)}$$

$$= \dfrac{x - 10}{(x+2)(x-1)}$$

> Simplify the numerator.

3 Express each of the following as a fraction in its simplest form.

a $\dfrac{2x}{3} + \dfrac{4x}{5}$

b $\dfrac{x+1}{3} + \dfrac{x+3}{2}$

c $\dfrac{2x-3}{2} + \dfrac{5x-1}{3}$

d $\dfrac{3}{x} + \dfrac{2}{x^2}, x \neq 0$

e $\dfrac{5}{2x} + \dfrac{6}{x^2}, x \neq 0$

f $\dfrac{4}{x+1} + \dfrac{3}{x-2}, x \neq -1, x \neq 2$

g $\dfrac{6}{p} + \dfrac{2}{p-2}, p \neq 0, p \neq 2$

h $\dfrac{x}{x-3} + \dfrac{3}{x+1}, x \neq 3, x \neq -1$

i $\dfrac{x+2}{x-1} + \dfrac{x-2}{x+3}, x \neq -3, x \neq 1$

4 Express each of the following as a fraction in its simplest form.

a $\dfrac{3x}{4} - \dfrac{2x}{5}$

b $\dfrac{x+2}{2} - \dfrac{x+1}{5}$

c $\dfrac{4x-1}{2} - \dfrac{2x-4}{3}$

d $\dfrac{5}{x} - \dfrac{4}{x^2}, x \neq 0$

e $\dfrac{7}{3x} - \dfrac{1}{2x^2}, x \neq 0$

f $\dfrac{5}{x+2} - \dfrac{2}{x-3}, x \neq 3, x \neq -2$

g $\dfrac{4}{y-2} - \dfrac{5}{y}, y \neq 2, y \neq 0$

h $\dfrac{4x}{x-3} - \dfrac{2}{x-2}, x \neq 3, x \neq 2$

i $\dfrac{x-5}{x-3} - \dfrac{x-1}{x+4}, x \neq 3, x \neq -4$

8 Determining the gradient of a straight line, given two points

Exercise 8A Gradient of a straight line

1 Find the gradient of each line.

a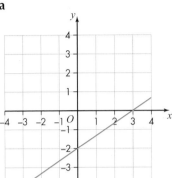

b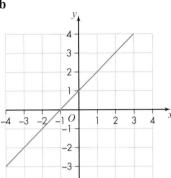

c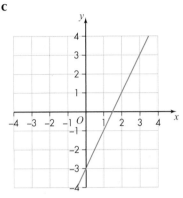

d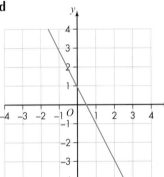

e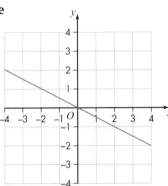

f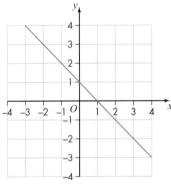

2 Calculate the gradient of the line joining each pair of points.

a (3, 5) and (4, 7)   b (5, 9) and (7, 17)   c (4, 6) and (5, 7)

d (1, 4) and (4, 19)   e (0, 11) and (4, 23)   f (2, 5) and (5, 5)

g (2, 5) and (3, −3)   h (2, 8) and (3, 2)   i (4, 8) and (8, 8)

j (8, 15) and (6, 33)   k (7, 12) and (4, 42)   l (4, 8) and (3, 14)

3 Calculate the gradient of the line joining each pair of points.

a (−1, 8) and (3, −4)   b (−4, −3) and (4, 1)   c (−3, 6) and (9, −2)

d (−9, −5) and (−3, −3)   e (−10, 2) and (−5, 12)   f (−5, 20) and (11, −4)

g (−2, 5.5) and (−5, 1.5)   h $\left(\frac{1}{2}, -\frac{1}{4}\right)$ and $\left(-\frac{1}{2}, \frac{3}{4}\right)$ i $\left(\frac{2}{3}, -\frac{1}{5}\right)$ and $\left(-\frac{3}{5}, \frac{1}{3}\right)$

4 a Calculate the gradient of the line joining point A (2, −3) and point B (−2, 7).

   b State the gradient of a line parallel to AB.

5 A is the point (1, 5) and B is the point (4, y). If the line joining points A and B has a gradient of 2, find the value of y.

6 A local council requires a disability access ramp to have a maximum gradient of $\frac{1}{15}$.

This ramp is designed to be built onto a library.

Does the ramp meet the requirements?
Give a reason for your answer.

0.6 m

12 m

9 Calculating the length of an arc or the area of a sector of a circle

Example

Calculate the length of the arc of this sector. Give your answer correct to 3 significant figures (3 s.f.).

55°
8 cm

Arc length = $\dfrac{\text{angle}}{360} \times \pi d$

$= \dfrac{55}{360} \times \pi \times 16$

$= 7.67944...$ ●————————(Using the π button on the calculator.)

$= 7.68\,\text{cm}$ (3 s.f.)

Example

Calculate the area of this sector. Take $\pi = 3.14$.

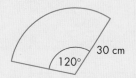
30 cm
120°

Sector area = $\dfrac{\text{angle}}{360} \times \pi r^2$

$= \dfrac{120}{360} \times 3.14 \times 30^2$

$= \dfrac{1}{3} \times 3.14 \times 900$ ●————————(Simplify fraction and calculate 30^2.)

$= \dfrac{1}{3} \times 900 \times 3.14$ ●————————(Rearrange.)

$= 300 \times 3.14 = 3 \times 100 \times 3.14 = 3 \times 314$

$= 942\,\text{cm}^2$

Exercise 9A Arc length and sector area

For each of the questions below, give your answer correct to 3 significant figures unless otherwise stated.

1 For each sector, calculate the length of the arc.

a

2 cm
80°

b

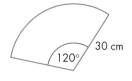
30 cm
120°

c

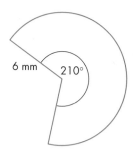
6 mm
210°

9 Calculating the length of an arc or the area of a sector of a circle

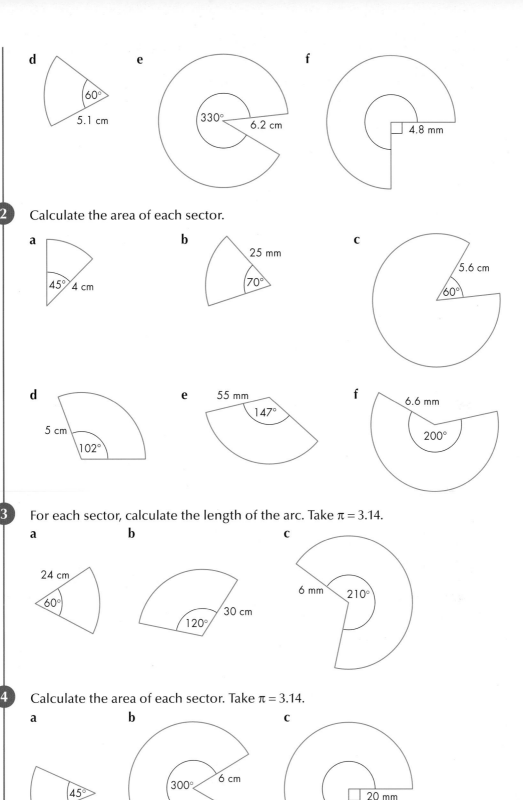

d

60°

5.1 cm

e

330°

6.2 cm

f

4.8 mm

2 Calculate the area of each sector.

a

45° 4 cm

b

25 mm

70°

c

5.6 cm

60°

d

5 cm

102°

e

55 mm

147°

f

6.6 mm

200°

3 For each sector, calculate the length of the arc. Take π = 3.14.

a

24 cm

60°

b

30 cm

120°

c

6 mm 210°

4 Calculate the area of each sector. Take π = 3.14.

a

45°

40 cm

b

300° 6 cm

c

20 mm

5

a Calculate the area of the shaded region of the shape.

b Calculate the perimeter of the shaded region of the shape.

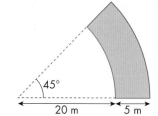

45°

20 m 5 m

6

a The area of this sector is 39 cm².

Find the value of x, the angle at the centre of the sector. Give your answer to the nearest degree.

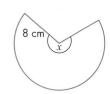

8 cm

x

b The length of the arc of this sector is 53.4 cm.

Find the value of x, the angle at the centre of the sector. Give your answer to the nearest degree.

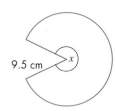

9.5 cm x

7

a The length of the arc of this sector is 10.5 cm.

Find the length of the radius.

50°

b The area of this sector is 398 mm².

Find the length of the radius.

8

There is an infrared sensor in a security system. The sensor can detect movement inside a sector of a circle. The radius of the circle is 16 m. The sector angle is 120°.

Calculate the area of the sector.

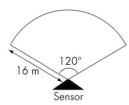

120°

16 m

Sensor

9

A circle of radius 8 cm is cut up into five congruent sectors.

Calculate the perimeter of each sector.

10

A shelf to fit in the corner of a room is to be cut in the shape of a quarter of a circle.

It will be cut to the maximum size from a square of wood of side 30 cm.

What will be the area of the shelf?

11

ABCD is a square of side length 15 cm. *APC* and *AQC* are arcs of circles with centres *D* and *B*.

Calculate the area of the unshaded part.

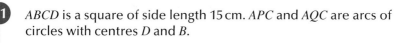

A B

P

Q

D C

10 Calculating the volume of a standard solid

Exercise 10A Volume of a solid

1 Calculate the volume of each sphere.

a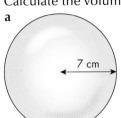
7 cm

b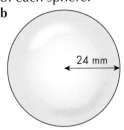
24 mm

c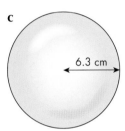
6.3 cm

d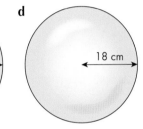
18 cm

> **Hint** Volume of a sphere, $V = \frac{4}{3}\pi r^3$.
> This is given in the exam.

2 Calculate the volume of each cone.

a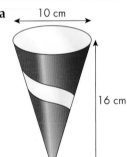
10 cm
16 cm

b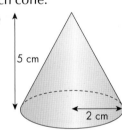
5 cm
2 cm

c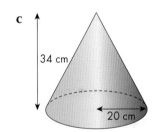
34 cm
20 cm

> **Hint** Volume of a cone, $V = \frac{1}{3}\pi r^2 h$.
> This is given in the exam.

3 Calculate the volume of each rectangular-based pyramid.

a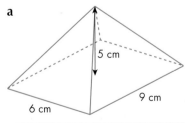
5 cm
9 cm
6 cm

b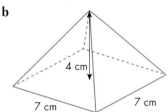
4 cm
7 cm
7 cm

> **Hint** Volume of a rectangular-based pyramid, $V = \frac{1}{3}Ah$.
> This is given in the exam.

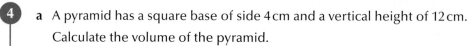

4 **a** A pyramid has a square base of side 4 cm and a vertical height of 12 cm. Calculate the volume of the pyramid.

b A pyramid has a square base of side 12 cm and a vertical height of 8 cm. Calculate the volume of the pyramid.

5 Calculate the volume of each sphere. Take π = 3.14.

a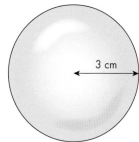

b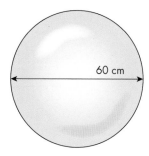

6 Calculate the volume of each cone. Take π = 3.14.

a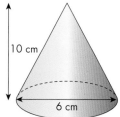

b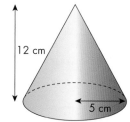

c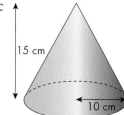

7 Calculate the volume of each rectangular-based pyramid.

a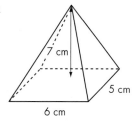

b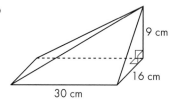

8 This shape consists of a cuboid and a rectangular-based pyramid. Calculate the volume of this shape.

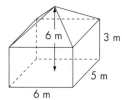

9 The diagram shows a spinning top made from a cone and a hemisphere. The radius of the sphere is 6 cm and the height of the spinning top is 17 cm.

Calculate the volume of the spinning top. Give your answer correct to 3 significant figures (3 s.f.).

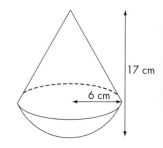

10 The pyramid in the diagram has its top 6 cm cut off. The remaining shape is called a frustum.

Calculate the volume of the frustum.

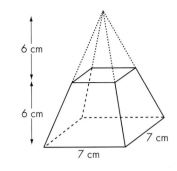

6 cm

6 cm

7 cm

7 cm

11 Calculate the height h of a rectangular-based pyramid with a length of 14 cm, a width of 10 cm and a volume of 140 cm³.

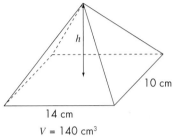

h

10 cm

14 cm

$V = 140$ cm³

12 The volume of a sphere is 50 m³.

Find its diameter.

13 An ice cream company sells ice cream in cones and pyramids.

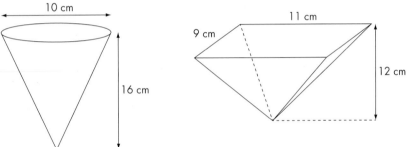

10 cm

16 cm

11 cm

9 cm

12 cm

The cone has a diameter of 10 cm and a height of 16 cm.

The pyramid has a rectangular base with length 11 cm and breadth 9 cm, and a height of 12 cm.

Which cone holds more ice cream? Justify your answer.

14 a A candle is in the shape of a cone with diameter 13 cm and height 15 cm.

Find the volume of the cone. Give your answer correct to 2 s.f.

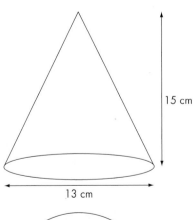

15 cm

13 cm

  b A second candle is in the shape of a hemisphere. It has the same volume as the cone in part **a**.

What is the radius of the hemisphere?

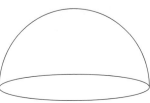

11 Rounding to a given number of significant figures

Exercise 11A Rounding to significant figures

1 Round each number to 1 significant figure.

    a 51 203   **b** 56 189   **c** 33 261   **d** 89 998   **e** 94 999

    f 53.71   **g** 87.24   **h** 31.06   **i** 97.835   **j** 184.23

2 Round each number to 2 significant figures.

    a 6725   **b** 35 724   **c** 68 522   **d** 41 689   **e** 27 308

    f 6973   **g** 2174   **h** 958   **i** 439   **j** 327.6

3 Round each number to the number of significant figures (s.f.) indicated.

    a 46 302 (1 s.f.)   **b** 6177 (2 s.f.)   **c** 89.67 (3 s.f.)   **d** 216.7 (2 s.f.)   **e** 7.78 (1 s.f.)

    f 1.087 (2 s.f.)   **g** 729.9 (3 s.f.)   **h** 5821 (1 s.f.)   **i** 66.51 (2 s.f.)   **j** 5.986 (1 s.f.)

    k 7.552 (1 s.f.)   **l** 9.7454 (3 s.f.)   **m** 25.76 (2 s.f.)   **n** 28.53 (1 s.f.)   **o** 869.89 (3 s.f.)

    p 35.88 (1 s.f.)   **q** 0.084 71 (2 s.f.)   **r** 0.0099 (2 s.f.)   **s** 0.0809 (1 s.f.)   **t** 0.061 97 (3 s.f.)

4 Write down the lower and upper bounds of each measurement. Each is given to the accuracy stated.

    a 6 m (1 s.f.)   **b** 34 kg (2 s.f.)   **c** 56 min (2 s.f.)   **d** 80 g (2 s.f.)

    e 3.70 m (2 d.p.)   **f** 0.9 kg (1 d.p.)   **g** 0.08 s (2 d.p.)   **h** 900 g (2 s.f.)

    i 0.70 m (2 d.p.)   **j** 360 ml (3 s.f.)   **k** 17 weeks (2 s.f.)   **l** 200 g (2 s.f.)

5 What are the least and the greatest possible numbers of people living in these towns?

    a Hellaby   population 900 (1 s.f.)

    b Hook   population 650 (2 s.f.)

    c Hundleton   population 1050 (3 s.f.)

6 A parking space is 4.8 m long, measured to the nearest tenth of a metre.

A car is 4.5 metres long, measured to the nearest half a metre.

Which of the following statements is definitely true?

A: The space is big enough.

B: The space is not big enough.

C: It is impossible to tell whether or not the space is big enough.

Explain how you decide.

7 Natasha has 20 identical bricks. Each brick is 15 cm long, measured to the nearest centimetre.

    a What is the greatest possible length of one brick?

    b What is the smallest possible length of one brick?

    c If the bricks are put end to end, what is the greatest possible length of all the bricks?

    d If the bricks are put end to end, what is the least possible length of all the bricks?

12 Determining the equation of a straight line, given the gradient

Example

Find the equation of the line passing through the points $(-2, 1)$ and $(6, 5)$.

$$m = \frac{5-1}{6-(-2)} = \frac{4}{8} = \frac{1}{2}$$

> Find the gradient between the two points. Make sure your answer is in its simplest form.

There are two methods for finding the equation once you have worked out the gradient.

Method 1: Use $y - b = m(x - a)$

Use $m = \frac{1}{2}$ and $(a, b) = (6, 5)$

> Either pair of coordinates can be used, but it is best to use the more straightforward coordinate.

$$y - 5 = \frac{1}{2}(x - 6)$$

$$y - 5 = \frac{1}{2}x - 3$$

> Substitute into the formula.

$$y = \frac{1}{2}x + 2$$

Method 2: Substitute into $y = mx + c$

$$y = \frac{1}{2}x + c$$

> Substitute the gradient into the formula.

$$5 = \frac{1}{2} \times 6 + c$$

> Substitute $(6, 5)$ for x and y.

$$5 = 3 + c$$

$$c = 2$$

$$y = \frac{1}{2}x + 2$$

Exercise 12A Equation of a straight line

1 Draw the following lines. Use the same grid, from −10 to 10 on both x- and y-axes.

a $y = 4x - 3$ **b** $y = 2x + 3$ **c** $y = -5x + 2$

d $y = 3x + 1$ **e** $y = x + 3$ **f** $y = -2x - 3$

2 Find the equation of each line, expressing your answers in the form $y = mx + c$.

a

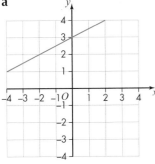

b

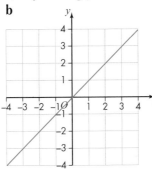

c

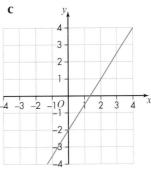

d

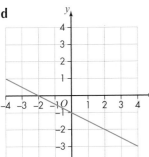

e

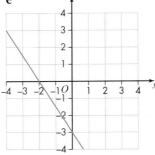

3 Write down **i** the gradient and **ii** the y-intercept of each line.

a $y = 4x + 3$ **b** $y = 3x - 2$ **c** $y = 2x + 1$

d $y = -3x + 3$ **e** $y = 5x$ **f** $y = -2x + 3$

g $y = x$ **h** $y = -\frac{1}{2}x + 3$ **i** $y = \frac{1}{4}x + 2$

4 Write down the equation of each line.

a parallel to $y = 3x - 2$ and passes through $(0, 4)$

b parallel to $y = \frac{1}{4}x + 3$ and passes through $(0, -1)$

c parallel to $y = -x + 3$ and passes through $(0, 2)$

5 Find the equation of each line. Express your answer in the form $y = mx + c$.

a gradient is -4; y-intercept is 2

b gradient is 3; line passes through the point $(6, 4)$

c gradient is 5; line passes through $(2, 7)$

d gradient is -2; line passes through $(4, -1)$

e gradient is 3; line passes through $(-4, -2)$

f gradient is $-\frac{1}{3}$; line passes through $(-3, 5)$

6 Find the equation of each line. Express your answer in the form $y = mx + c$.

a line passes through the points $(2, 18)$ and $(5, 9)$

b line passes through the points $(-6, -7)$ and $(18, 19)$

c line passes through the points $(-3, 1)$ and $(2, 11)$

d line passes through the points $(-3, 14)$ and $(3, -4)$

e line passes through the points $(-8, -6)$ and $(2, -1)$

f line passes through the points $(-5, 7)$ and $(-1, -5)$

g line passes through the points $(-6, -7)$ and $(3, -1)$

h line passes through the points $(-5, -3)$ and $(7, 6)$

7 The line segment AB joins A $(1, 4)$ and B $(5, 2)$. Find the equation of the line that passes through $(4, 7)$ and is parallel to AB.

8 For each equation:

i rearrange the equation in the form $y = mx + c$

ii state the gradient of the line

iii state the coordinates of the point of intersection with the y-axis.

a $3y = 4x + 6$ **b** $2y = -5x - 4$

c $8y = 4x - 12$ **d** $x - y = 0$

e $y - 7 = 3x$ **f** $y - 2x - 4x = 0$

g $6x + 4y = 12$ **h** $2x + 3y = 12$

i $4x - 5y = 40$ **j** $x + y = 6$

k $3x - 2y = 24$ **l** $x - y = -6$

Example

A function has equation $f(x) = 3x - 7$.

a Find:

   i $f(4)$

   ii $f(-2)$

b Given that $f(t) = -9$, calculate the value of t.

a **(i)** $f(4) = 3 \times 4 - 7 = 12 - 7 = 5$ ($f(4)$ means substitute 4 in place of x.)

   (ii) $f(-2) = 3 \times (-2) - 7 = -6 - 7 = -13$

b $f(t) = 3t - 7 = -9$ (Substitute t in place of x. The solution is -9.)

   $3t - 7 = -9$ (Now solve this equation.)

     $3t = -2$

       $t = -\frac{2}{3}$

Exercise 12B Functions

1 A function has equation $f(x) = 5x + 14$.

   a Find:

      i $f(7)$ **ii** $f(-2)$ **iii** $f(0)$ **iv** $f(-12)$

   b Given than $f(t) = 16$, calculate the value of t.

   c Given than $f(p) = 2$, calculate the value of p.

2 A function has equation $f(x) = -\frac{1}{2}x - 8$.

   a Find:

      i $f(4)$ **ii** $f(-6)$ **iii** $f(1)$ **iv** $f(-18)$

   b Given than $f(t) = -6$, calculate the value of t.

   c Given than $f(p) = 3\frac{1}{2}$, calculate the value of p.

3 $f(x) = 3x^2 - 2$

   a Find the value of:

      i $f(2)$ **ii** $f(5)$ **iii** $f(-1)$ **iv** $f(-4)$ **v** $f(\sqrt{3})$

   b Given that $f(k) = 25$, find both values of k.

> **Hint** Remember that the square root of a number has a positive solution and a negative solution.

4 $g(x) = 8 - x^2$

   a Find the value of:

      i $g(2)$ **ii** $g(-3)$ **iii** $g(6)$ **iv** $g(-4)$ **v** $g(\sqrt{7})$ **vi** $g(-0.5)$

   b Solve $g(x) = -1$.

5 $f(x) = 2x^2 - 6x + 4$

   a Find the value of:

      i $f(-1)$ **ii** $f(5)$ **iii** $f(-2)$

   b Solve $f(x) = 0$.

> **Hint** You need to solve a quadratic equation for this question. If you need help on solving these equations, see Chapter 19 in the *National 5 Maths Student Book*.

13 Working with linear equations and inequations

Example

Solve the equation for x.

$3(2x + 1) + 2(4x - 5) = 4(3x - 7)$

$6x + 3 + 8x - 10 = 12x - 28$ ———• (Multiply out and remove the brackets.)

$14x - 7 = 12x - 28$ ———• (Simplify the left-hand side of the equation.)

$2x = -21$

$x = -\frac{21}{2}$ ———• (The answer should always be in its simplest form. Improper fractions or mixed number answers are accepted.)

Exercise 13A Solving linear equations

1 Solve these equations.

   a $2x + 1 = x + 3$ **b** $3y + 2 = 2y + 6$ **c** $5a - 3 = 4a + 4$

   d $5t + 3 = 3t + 9$ **e** $7p - 5 = 5p + 3$ **f** $6k + 5 = 3k + 20$

   g $6m + 1 = m + 11$ **h** $5s - 1 = 2s - 7$ **i** $4w + 8 = 2w + 8$

2 Solve these equations.

   a $2(x + 1) = 8$ **b** $3(x - 3) = 12$

   c $3(t + 2) = 9$ **d** $2(x + 5) = 20$

   e $2(2y - 5) = 14$ **f** $2(3x + 4) = 26$

   g $4(3t - 1) = 20$ **h** $2(t + 5) = 6$

3 Solve these equations.

   a $5(t - 2) = 4t - 1$ **b** $4(x + 2) = 2(x + 1)$

   c $5p - 2 = 5 - 2p$ **d** $2(2x + 3) = 3(x - 4)$

4 Solve these equations.

   a $3(2b - 1) + 25 = 4(3b + 1)$ **b** $3(4c + 1) - 17 = 2(3c + 2) - 3c$

5 Solve the equation $(x - 2)(x + 3) = (x + 5)(x - 3)$.

 6 **a** Show that the equation $4(3x + 5) = 3(4x + 2)$ cannot be solved.

   b Show that there are an infinite number of solutions to the equation
      $5(2x + 8) = 2(5x + 20)$.

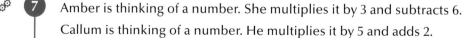 **7** Amber is thinking of a number. She multiplies it by 3 and subtracts 6.

   Callum is thinking of a number. He multiplies it by 5 and adds 2.

   Amber and Callum discover that they both thought of the same original number and
   both got the same final answer.

   What number did they think of?

8 On a bookshelf there are $2b$ crime novels, $3b - 2$ science fiction novels and $b + 7$ romance novels. There are 65 books altogether.

How many of each type of book are there?

9 The triangle shown is isosceles.

What is the perimeter of the triangle?

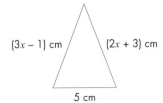

$(3x - 1)$ cm $(2x + 3)$ cm

5 cm

10 Could this triangle be equilateral? Explain your answer.

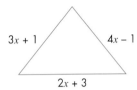

$3x + 1$ $4x - 1$

$2x + 3$

11 Solve these equations.

   a $\dfrac{g}{3} + 2 = 8$ **b** $\dfrac{m}{4} - 5 = 2$ **c** $\dfrac{h}{8} - 3 = 5$ **d** $\dfrac{2h}{3} + 3 = 7$ **e** $\dfrac{3t}{4} - 3 = 6$

   f $\dfrac{3x}{4} - 1 = 8$ **g** $\dfrac{x + 5}{3} = 2$ **h** $\dfrac{t + 12}{2} = 5$ **i** $\dfrac{w - 3}{5} = 3$ **j** $\dfrac{y - 9}{2} = 3$

12 Solve these equations.

   a $\dfrac{2x - 1}{3} = 5$ **b** $\dfrac{5t - 4}{2} = 3$ **c** $\dfrac{4m + 1}{5} = 5$ **d** $\dfrac{8p - 6}{5} = 2$

   e $\dfrac{5x + 1}{4} = 4$ **f** $\dfrac{17 + 2t}{9} = 1$ **g** $\dfrac{2 + 4x}{3} = 4$ **h** $\dfrac{8 - 2x}{11} = 1$

Example

Solve the equation for x.

$$\dfrac{3x}{4} - \dfrac{1}{8} = 2x$$

$\dfrac{3x}{4} (\times 8) - \dfrac{1}{8} (\times 8) = 2x (\times 8)$ ●——— Multiply each term by 8, which is the lowest common multiple of 4 and 8, the denominators of the fractions.

$6x - 1 = 16x$

$-10x = 1$ ●——— Rearrange.

$x = -\dfrac{1}{10}$ ●——— Solve, giving your answer in its simplest form.

13 Solve these equations.

   a $\dfrac{4x}{3} - \dfrac{1}{6} = 3x$ **b** $\dfrac{5x}{4} + \dfrac{3}{8} = 2x$ **c** $\dfrac{2x}{5} - \dfrac{3}{4} = 4x$ **d** $\dfrac{5x}{3} - 7 = \dfrac{2x}{5}$

   e $\dfrac{2x}{3} + 5 = \dfrac{3x}{7}$ **f** $\dfrac{5x}{6} - \dfrac{2}{3} = \dfrac{7x}{2}$ **g** $\dfrac{x + 3}{2} + \dfrac{x - 1}{3} = 5$ **h** $\dfrac{x - 2}{4} - \dfrac{x - 3}{3} = \dfrac{3}{2}$

Exercise 13B Solving linear inequations

  1 Draw diagrams to illustrate these inequalities.

    a $x \leqslant 2$ **b** $x > -3$ **c** $x \geqslant 1$ **d** $x < 4$

    e $x \geqslant -3$ **f** $1 < x \leqslant 4$ **g** $-2 \leqslant x \leqslant 4$ **h** $-2 < x < 3$

  2 Solve the following linear inequalities.

    a $x + 5 \geqslant 9$ **b** $x + 4 < 2$

    c $x - 2 \leqslant 3$ **d** $x - 5 > -2$

    e $4x + 3 \leqslant 9$ **f** $5x - 4 \geqslant 16$

    g $2x - 1 > 13$ **h** $3x + 6 \leqslant 3$

3 Solve the following linear inequalities.

    a $2(x - 3) < 14$ **b** $4(3x + 2) < 32$

    c $3x + 1 > 2x - 5$ **d** $6t - 5 < 4t + 3$

    e $2y - 11 < y - 5$ **f** $3x + 2 > x + 3$

    g $4w - 5 < 2w + 2$ **h** $2(5x - 1) < 2x + 3$

Example

Solve the linear inequality for x.

$9 - 2(3x - 5) > 14$

$9 - 2(3x - 5) > 14$ •————————— Multiply out and remove the brackets.

$9 - 6x + 10 > 14$ •————————— Simplify.

$19 - 6x > 14$ •————————— Rearrange.

$-6x > -5$ •————————— Divide both sides of the inequation by -6. Remember to change the

$x < \frac{5}{6}$ direction of the inequality sign when dividing by a negative number.

  4 Solve these linear inequalities.

    a $5 - 3(2 + x) < 12$ **b** $6 + 4(3x - 5) > 11$

    c $12 - 2(5x + 1) \leqslant 12$ **d** $15 - 2(3x - 4) \geqslant 5x$

    e $8 - 6(3x - 1) < 7$ **f** $3(2x - 3) + 4(x - 1) > 8$

    g $4(x - 7) - 3(3x - 1) \leqslant 6$ **h** $4(2x - 5) - 7 > 6(2x - 1)$

5 Solve the following inequalities and illustrate their solutions on number lines.

    a $\dfrac{5x + 2}{2} > 3$ **b** $\dfrac{3x - 4}{5} \leqslant 1$

    c $\dfrac{4x + 3}{2} \geqslant 11$ **d** $\dfrac{2x - 5}{4} < 2$

    e $\dfrac{8x + 2}{3} \leqslant 2$ **f** $\dfrac{7x + 9}{5} > -1$

    g $\dfrac{x - 2}{3} \geqslant -3$ **h** $\dfrac{5x - 2}{4} \leqslant -1$

  6 Solve these linear inequalities.

    a $\dfrac{4x}{5} - \dfrac{3}{10} > 2$ **b** $\dfrac{5x}{6} + \dfrac{2}{3} < 4$

    c $\dfrac{3x}{8} - \dfrac{1}{4} \geqslant \dfrac{x}{2}$ **d** $\dfrac{x - 3}{2} - \dfrac{x - 1}{5} \leqslant 6$

14 Working with simultaneous equations

Exercise 14A Solving simultaneous equations graphically and algebraically

1 For each question, draw a graph to solve the pair of simultaneous equations.

a $y = 3x - 1$
$y = 2x$

b $y = 2x - 1$
$y = x$

c $y = 3x - 2$
$y = x - 2$

d $y = 3 - 2x$
$y = x$

e $-x + y = 5$
$y = 2x - 1$

f $2x + y = 6$
$x + y = -6$

g $x - y = 3$
$x + y = 5$

h $x + y = -5$
$y = 4x$

i $x + 4y = 1$
$x - y = 6$

j $y = 2x + 1$
$3x + 2y = 23$

k $y = 2x + 5$
$y = x + 4$

l $y = x$
$x - y = 4$

2 Solve each pair of simultaneous equations.

a $x + 3y = 11$
$x + 2y = 9$

b $3x + 4y = 25$
$3x + 2y = 17$

c $3x - 2y = 8$
$4x + 2y = 20$

3 Solve each pair of simultaneous equations.

a $3x + 2y = 12$
$4x - y = 5$

b $4x + 3y = 37$
$2x + y = 17$

c $2x + 3y = 19$
$6x + 2y = 22$

d $5x - 2y = 14$
$3x - y = 9$

4 Solve each pair of simultaneous equations.

a $6x + 5y = 23$
$5x + 3y = 18$

b $3x - 4y = 13$
$2x + 3y = 20$

c $8x - 2y = 14$
$6x + 4y = 27$

d $5x + 2y = 33$
$4x + 5y = 23$

5 Solve each pair of simultaneous equations.

a $2x + 4y = 24$
$y = x + 3$

b $3x + y = 22$
$y = 5x - 26$

c $6x - 3y = 18$
$y = x - 2$

6 Solve each pair of simultaneous equations.

a $3x + 4y = 19$
$y = 6 - 2x$

b $6x - 2y = 24$
$x = 20 - 5y$

c $2x - y = 10$
$x = 4y - 2$

7 Solve each pair of simultaneous equations.

a $6x + 10y = 72$
$y = 31 - 4x$

b $5x + y = 14$
$5x - 6y = -14$

For each of the following questions:

- read each situation carefully
- write a pair of simultaneous equations and use this to solve the problem
- answer the question within the context of the problem.

8 A book and a CD cost £14.00 together. The CD costs £7.00 more than the book.

How much does each cost?

9 It costs two adults and three children £28.50 to go to the cinema.
It costs three adults and two children £31.50 to go to the cinema.

Let the price of an adult ticket be £x and the price of a child ticket be £y.

Find the price of each type of ticket.

10 Ina wants to buy some snacks for her friends. She works out from the labels that two cakes and three bags of peanuts contain 63 g of fat and that one cake and four bags of peanuts contain 64 g of fat.

How many grams of fat are there in:

a one cake   **b** one bag of peanuts?

11 Ten second-class and six first-class stamps cost £4.96.
Eight second-class and ten first-class stamps cost £5.84.

How much would I pay for three second-class and four first-class stamps?

12 Two people bought cola and chocolate at the local store. Henri paid £4.37 for six cans of cola and five chocolate bars. Evie paid £2 for three cans of cola and two chocolate bars.

How much would it cost Mark to buy two cans of cola and a chocolate bar?

13 The difference between my son's age and my age is 28 years.
Five years ago my age was double that of my son.

Let my age now be x and my son's age now be y.

a Explain why $x - 5 = 2(y - 5)$.   **b** Find the values of x and y.

14 Four apples and two oranges cost £2.04.
Five apples and one orange cost £1.71.

Marcus buys four apples and eight oranges.

How much change will he get from a £10 note?

15 Five bags of compost and four bags of pebbles have a mass of 340 kg.
Three bags of compost and five bags of pebbles have a mass of 321 kg.

Carol needs six bags of compost and eight bags of pebbles.
Her trailer has a safe working load of 500 kg.

Can Carol transport all the bags safely on her trailer?

15 Changing the subject of a formula

Example

Change the subject to x.

$\frac{1}{2}ax^2 + t = S$

$\frac{1}{2}ax^2 = S - t$ — Subtract t from both sides.

$ax^2 = 2(S - t)$ — Multiply each side by 2.

$x^2 = \frac{2(S - t)}{a}$ — Divide each side by a to leave x^2 on the left-hand side.

$x = \sqrt{\frac{2(S - t)}{a}}$ — Take the square root of both sides.

Exercise 15A Changing the subject of a formula

1 Make x the subject of each formula.

  a $x + h = k$ **b** $p - x = m$ **c** $y = mx + c$

  d $k + 2bx = 9b$ **e** $A = fgx$ **f** $N = 3xy$

2 Make x the subject of each formula.

  a $y = \frac{4x}{3}$ **b** $p = \frac{3x - 5}{2}$ **c** $c = \frac{x}{3} - g$

  d $M = \frac{nx}{b}$ **e** $R = \frac{p - x}{h}$ **f** $L = \frac{H(x - j)}{g}$

3 Change the subject of the formula to the letter in brackets.

  a $y = \frac{7}{t}$ (t) **b** $d = \frac{c}{h + 2t}$ (h) **c** $M = \frac{J}{7w} + k$ (w)

4 Make the letter in brackets the subject of the formula.

  a $4(x - 2y) = 3(2x - y)$ (x) **b** $p(a - b) = q(a + b)$ (a) **c** $A = 2ab^2 + ac$ (a)

  d $s(t + 1) = 2r + 3$ (r) **e** $st - r = 2r - 3t$ (t) **f** $f(g - 2x) = Ag + t$ (g)

5 Make x the subject of each formula.

  a $ax = b - cx$ **b** $x(a - b) = x + b$ **c** $a - bx = dx - a$

  d $x(c - d) = c(d - x)$ **e** $x(a - 5) = 2(b + x)$ **f** $\frac{x(b + c)}{2} = b(x + c)$

6 Make x the subject of each formula.

  a $S = \sqrt{3x + y}$ **b** $B = hx^2 + j$ **c** $K = (x + 3)^2$

  d $G = \sqrt{\frac{hx}{m}}$ **e** $H = cf + \frac{1}{2}xt^2$ **f** $C = \frac{k}{\sqrt{x - p}}$

  g $L = \left(\frac{k + 2x}{t}\right)^2$ **h** $P = j\sqrt{\frac{r}{x - a}}$

Exercise 16A Equations and graphs of quadratic functions

1. The equation of each of the graphs below is $y = kx^2$.

   For each of the graphs, find the value of k.

   a
   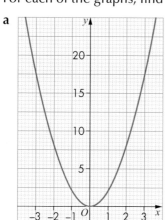

   b
   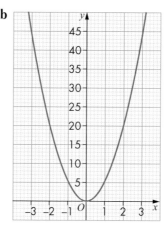

   c
   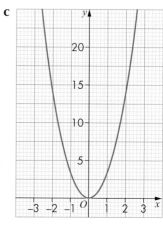

   d
   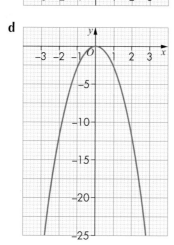

   e
   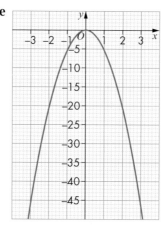

   f
   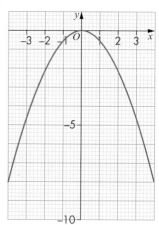

2. The equation of each of the graphs below is $y = kx^2$.

   For each of the graphs, find the value of k.

   a   **b**   **c**
   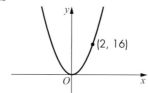   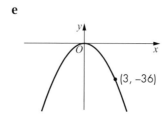

   (4, 48)   (2, 16)   (−3, 18)

   d   **e**   **f**
   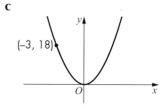   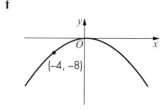

   (2, −20)   (3, −36)   (−4, −8)

3 The equation of each of the graphs below is $f(x) = (x + a)^2 + b$.

For each of the graphs, find the values of a and b.

a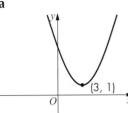

(3, 1)

b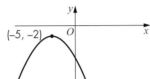

(4, 2)

c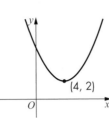

(−3, 2)

4 The equation of each of the graphs below is $g(x) = -(x + a)^2 + b$.

For each of the graphs, find the values of a and b.

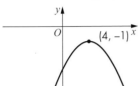

(4, −1)

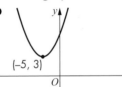

(−5, −2)

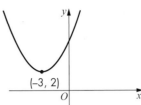

(−4, 2)

5 The equation of each of the graphs below can be written as $f(x) = (x + a)^2 + b$ or $f(x) = -(x + a)^2 + b$.

For each graph:

i identify the values of a and b

ii write down the equation of the graph.

a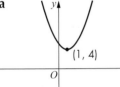

(1, 4)

b

(−5, 3)

c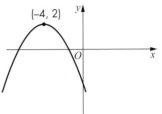

(−3, −1)

d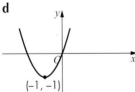

(−1, −1)

e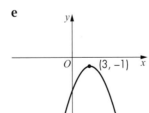

(3, −1)

f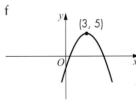

(3, 5)

6 The equation of this graph can be written as $f(x) = (x + a)^2 + b$.

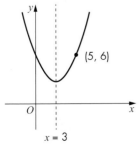

(5, 6)

$x = 3$

Find the values of a and b.

17 Sketching a quadratic function

Example

Sketch the graph of $y = (x - 3)(x + 1)$.

Roots

$(x - 3)(x + 1) = 0$ 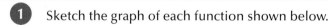 ┤ Graph cuts the x-axis when $y = 0$.

$x - 3 = 0$ or $x + 1 = 0$

$x = 3$ or $x = -1$

Roots are $(3, 0)$ and $(-1, 0)$.

y-intercept

$y = (0 - 3)(0 + 1)$ ┤ Graph cuts the y-axis when $x = 0$.

$\quad = (-3) \times 1 = -3$

y-intercept has coordinates $(0, -3)$

Turning point

$x = \dfrac{3 + (-1)}{2} = 1$ ┤ The x-coordinate of the turning point is halfway between the roots.

$y = (1 - 3)(1 + 1)$ ┤ Substitute the x-coordinate into the original equation to find the y-coordinate.

$y = (-2) \times 2 = -4$

Turning point has coordinates $(1, -4)$

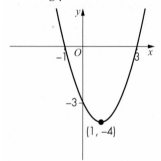

Exercise 17A Sketching a quadratic function

1 Sketch the graph of each function shown below.

On your sketch, show clearly the coordinates of the turning point and the points of intersection with the x- and y- axes.

   a $y = (x - 2)(x - 6)$ **b** $y = (x - 3)(x + 5)$ **c** $y = (x - 4)(x + 8)$

   d $f(x) = (x - 5)(x - 1)$ **e** $f(x) = (x - 6)(x + 4)$ **f** $f(x) = (x - 7)(x + 3)$

   g $g(x) = x(x - 4)$ **h** $y = (6 - x)(x + 2)$

2 Sketch the graph of each function shown below.

On your sketch, show clearly the coordinates of the turning point and the points of intersection with the x- and y- axes.

   a $y = x^2 + 4x - 5$ **b** $y = x^2 - 9$ **c** $y = 6x - x^2$

   d $f(x) = x^2 + 2x - 8$ **e** $f(x) = 32 + 4x - x^2$ **f** $f(x) = x^2 + 8x + 7$

   g $g(x) = 16 - x^2$ **h** $y = x^2 + x - 6$

3 Sketch the graph of each function shown below.

On your sketch, show clearly the coordinates of the turning point and the point of intersection with the y-axis.

a $y = (x + 3)^2 + 1$ **b** $y = (x - 2)^2 + 3$ **c** $y = (x + 5)^2 + 4$

d $f(x) = (x + 2)^2 + 5$ **e** $f(x) = (x - 3)^2 + 2$ **f** $f(x) = (x - 1)^2 + 7$

g $g(x) = (x + 2)^2 + 6$ **h** $y = (x - 4)^2 - 2$

4 Sketch the graph of each function shown below.

On your sketch, show clearly the coordinates of the turning point and the point of intersection with the y-axis.

a $y = -(x + 2)^2 - 1$ **b** $y = -(x - 3)^2 - 2$ **c** $y = -(x - 5)^2 - 6$

d $f(x) = -(x + 3)^2 - 4$ **e** $f(x) = -(x + 4)^2 - 5$ **f** $f(x) = -(x - 2)^2 - 3$

g $g(x) = -(x + 1)^2 - 4$ **h** $y = -(x + 5)^2 + 3$

Example

Sketch the graph of $y = 3(x + 2)^2 + 4$.

Turning point

Coordinates of the turning point are (−2, 4).   ● ─────── State explicitly.

Minimum turning point   ● ─────── The coefficient of x^2 is positive.

y-intercept

$y = 3(0 + 2)^2 + 4$ ● ─────── Substitute $x = 0$ into the original equation.

   $= 3 \times 2^2 + 4$

   $= 3 \times 4 + 4$

   $= 16$

Coordinates of the y-intercept are (0, 16).

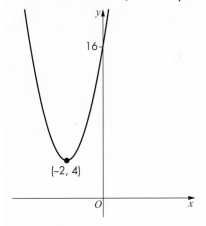

5 Sketch the graph of each function shown below.

On your sketch, show clearly the coordinates of the turning point and the point of intersection with the y-axis.

a $y = 2(x + 3)^2 + 2$ **b** $y = 3(x - 4)^2 + 1$ **c** $y = -2(x - 2)^2 - 4$

d $f(x) = -4(x + 1)^2 - 3$ **e** $f(x) = 5(x + 1)^2 + 4$ **f** $f(x) = -2(x - 3)^2 - 1$

g $g(x) = \frac{1}{2}(x + 2)^2 - 5$ **h** $y = -3(x + 1)^2 + 3$

18 Identifying features of a quadratic function

Example

A parabola has equation $y = (x - 2)^2 + 4$.

a State the equation of the axis of symmetry.

b State the coordinates of the turning point.

c State whether the graph has a minimum or maximum turning point.

a $x = 2$

From the equation $y = k(x + a)^2 + b$, the equation of the line of symmetry is $x = -a$ (which is the x-coordinate of the turning point).

b (2, 4)

From the equation $y = k(x + a)^2 + b$, the turning point is $(-a, b)$.

c Minimum

From the equation $y = k(x + a)^2 + b$, the graph has a minimum turning point if $k > 0$ and a maximum turning point if $k < 0$.

Exercise 18A Features of a quadratic function

1 For each quadratic function given below:

i state the equation of the axis of symmetry

ii state the coordinates of the turning point

iii state whether the graph has a minimum or maximum turning point.

a $y = (x - 3)^2 + 5$ **b** $y = (x + 2)^2 + 1$ **c** $y = -(x - 7)^2 - 1$

d $f(x) = -(x + 5)^2 - 8$ **e** $f(x) = (x - 8)^2 + 7$ **f** $g(x) = -(x + 2)^2 + 12$

g $y = (x + 3)^2 + 9$ **h** $f(x) = -(x + 6)^2 + 3$ **i** $y = (x + 7)^2 - 18$

2 For each quadratic function given below:

i state the equation of the axis of symmetry

ii state the coordinates of the turning point

iii state whether the graph has a minimum or maximum turning point.

a $y = 2(x - 4)^2 + 8$ **b** $y = 3(x + 1)^2 + 8$ **c** $y = -3(x - 6)^2 - 3$

d $f(x) = -4(x + 1)^2 - 5$ **e** $f(x) = 5(x - 8)^2 + 6$ **f** $g(x) = -2(x + 5)^2 + 12$

g $y = 6(x + 2)^2 + 14$ **h** $f(x) = -5(x + 4)^2 + 7$ **i** $y = 7(x + 3)^2 - 13$

3 For each of the quadratic functions given below:

i express the function in the form $y = (x + a)^2 + b$

> **Hint**   See Chapter 5 for help on completing the square.

ii state the equation of the axis of symmetry

iii state the coordinates of the turning point.

a $y = x^2 + 4x + 7$ **b** $y = x^2 + 6x + 11$ **c** $y = x^2 - 10x - 2$

d $y = x^2 - 2x - 7$ **e** $y = x^2 - 6x - 3$ **f** $y = x^2 + 12x + 35$

g $y = x^2 - 4x - 2$ **h** $y = x^2 - 14x + 53$ **i** $y = x^2 + x + 1$

4 Match each equation with the correct graph below.

   i $y = (x - 1)^2 + 3$ **ii** $y = -(x - 1)^2 + 3$ **iii** $y = 2(x - 1)^2 + 3$

   iv $y = (x + 1)^2 + 3$ **v** $y = -(x + 1)^2 + 3$ **vi** $y = 3(x + 1)^2 + 3$

a
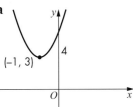

b
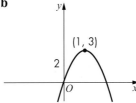

c
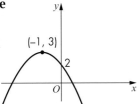

d
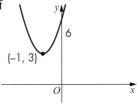

e
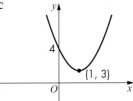

f
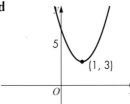

Exercise 18B Using quadratic functions to solve problems

1   The diagram below shows the path of a ball thrown into the air from ground level. The height (in metres) of the ball t seconds after being thrown is given by the formula $h(t) = 12t - t^2$.

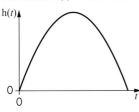

   a   What is the maximum height the ball reaches?

   b   How long does it take to reach the ground?

> Hint   For part **a**, find the y-coordinate of the turning point.
>
>   For part **b**, find the non-zero root of the function.

2 The diagram below shows the path of a distress flare fired into the air from ground level. The height (in metres) of the distress flare t seconds after being fired is given by the formula $h(t) = 42t - t^2$.

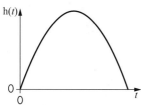

a How long does it take to reach the ground?

b Does the flare ever reach a height of 450 metres?
Give a reason for your answer.

> **Hint** Find the maximum height reached and compare your answer to 450 m.

3 A farmer has 20 metres of fence and wants to make a rectangular pen.

a If one side of the fence is x metres, express the length of the other side in terms of x.

b Show that the area A of the pen can be given as $A = 10x - x^2$.

c Find, algebraically, the length of x which would give the maximum area of the pen.

d What is the maximum area of the pen?

4 A theatre has 2000 seats. The profit P of a theatre ticket of £x is expressed as the formula $P = 2000x - 100x^2$.

a Find x, the price that should be charged for maximum profit.

b How much profit will the theatre make if it charges tickets at this price?

19 Working with quadratic equations

Exercise 19A Solving quadratic equations by factorising

1 Solve these equations.

a $(x + 3)(x + 2) = 0$ **b** $(t + 4)(t + 1) = 0$ **c** $(a + 5)(a + 3) = 0$

d $(x + 4)(x - 1) = 0$ **e** $(x + 2)(x - 5) = 0$ **f** $(t + 3)(t - 4) = 0$

g $(x - 2)(x + 1) = 0$ **h** $(x - 1)(x + 4) = 0$ **i** $(a - 6)(a + 5) = 0$

j $(x - 2)(x - 5) = 0$ **k** $(x - 2)(x - 1) = 0$ **l** $(a - 2)(a - 6) = 0$

2 Solve these equations.

a $x(x - 3) = 0$ **b** $x(x + 6) = 0$ **c** $(x + 3)(x - 3) = 0$

d $(4 + x)(5 - x) = 0$ **e** $(2x - 1)(x + 3) = 0$ **f** $(3x - 1)(2x + 5) = 0$

3 First factorise, then solve each of these equations.

a $x^2 + 6x + 5 = 0$ **b** $x^2 + 9x + 18 = 0$ **c** $x^2 - 7x - 8 = 0$

d $x^2 - 4x - 21 = 0$ **e** $x^2 + 3x - 10 = 0$ **f** $x^2 + 2x - 15 = 0$

g $t^2 - 4t - 12 = 0$ **h** $t^2 - 3t - 18 = 0$ **i** $x^2 + x - 2 = 0$

j $x^2 - 4x + 4 = 0$ **k** $m^2 - 10m + 25 = 0$ **l** $t^2 - 10t + 16 = 0$

m $t^2 + 7t + 12 = 0$ **n** $k^2 - 3k - 18 = 0$ **o** $a^2 - 20a + 64 = 0$

4 Solve these equations.

a $x^2 - 5x = 0$ **b** $x^2 + 8x = 0$ **c** $2x^2 - 14x = 0$

d $x^2 - 9 = 0$ **e** $x^2 - 25 = 0$ **f** $36 - x^2 = 0$

5 Solve these equations.

a $2x^2 + 5x + 2 = 0$ **b** $7x^2 + 8x + 1 = 0$ **c** $4x^2 + 3x - 7 = 0$

d $6x^2 + 13x + 5 = 0$ **e** $6x^2 + 7x + 2 = 0$ **f** $12x^2 - 25x + 12 = 0$

6 Rearrange these equations into the general form and then solve them.

a $x^2 - x = 6$ **b** $2x(4x + 7) = -3$ **c** $(x + 3)(x - 4) = 18$

d $11x = 21 - 2x^2$ **e** $(2x + 3)(2x - 3) = 9x$ **f** $x^2 = \dfrac{3 - x}{2}$

Exercise 19B Solving quadratic equations using the quadratic formula

1 Use the quadratic formula to solve these equations. Give your answers to 2 decimal places.

a $3x^2 + x - 5 = 0$ **b** $2x^2 + 4x + 1 = 0$

c $x^2 - x - 7 = 0$ **d** $3x^2 + x - 1 = 0$

e $3x^2 + 7x + 3 = 0$ **f** $2x^2 + 11x + 1 = 0$

g $2x^2 + 5x + 1 = 0$ **h** $x^2 + 2x - 9 = 0$

> **Hint** The quadratic formula for the equation $ax^2 + bx + c = 0$ is:
>
> $$x = \frac{-b \pm \sqrt{b^2 - 4ac}}{2a}$$

2 Use the quadratic formula to solve these equations. Give your answers correct to 3 significant figures.

a $x^2 + 5x + 3 = 0$

b $x^2 - 8x + 2 = 0$

c $x^2 + 7x - 4 = 0$

d $2x^2 + 4x + 1 = 0$

e $2x^2 - 6x + 3 = 0$

f $3x^2 - x - 5 = 0$

g $7 + 2x - x^2 = 0$

h $2x^2 = x + 5$

i $x^2 = \dfrac{x + 1}{5}$

3 Work out the discriminant $(b^2 - 4ac)$ of each function and determine the nature of its roots.

> **Hint** For $b^2 - 4ac > 0$, the function has two real, distinct roots.
>
> For $b^2 - 4ac = 0$, the function has two real, equal roots.
>
> For $b^2 - 4ac < 0$, the function has no real roots.

a $f(x) = x^2 + x + 3$

b $f(x) = 3x^2 - 4x + 2$

c $f(x) = x^2 - 6x - 12$

d $f(x) = 8x^2 + 8x + 2$

e $f(x) = 36x^2 - 9x$

f $f(x) = 4x^2 - 9$

g $f(x) = x^2 - 7x + 2$

h $f(x) = x^2 + 5x + 9$

i $f(x) = 2x^2 - 3x + 4$

j $f(x) = 4x^2 - 4x + 1$

k $f(x) = 5x^2 + 2x + 7$

l $f(x) = 8 - 3x - x^2$

4 Gerard is solving a quadratic equation using the formula.

He correctly substitutes values for a, b and c to get:

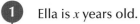

$$x = \frac{4 \pm \sqrt{112}}{6}$$

State a possible equation that Gerard is trying to solve.

5 Eric uses the quadratic formula to solve $9x^2 - 12x + 4 = 0$.

Anna uses factorisation to solve the same equation.

They both find something unusual in their solutions.

Explain what this is, and why.

Exercise 19C Using quadratic equations to solve problems

1 Ella is x years old.

Ella's brother is 4 years older than she is.

The product of their ages is 1020.

a State Ella's brother's age in terms of x.

b Show that $x^2 + 4x - 1020 = 0$.

c How old is Ella?

2 A right-angled triangle has sides $2x$ cm, $(2x + 1)$ cm, and $(x + 1)$ cm.

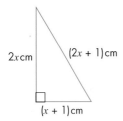

$2x$ cm $(2x + 1)$ cm

$(x + 1)$ cm

a Show that $x^2 - 2x = 0$.

b Find the value of x.

c Find the area of the triangle.

3 A rectangular lawn is 5 m longer than it is wide.

x m

length

a State the length of the lawn in terms of *x*.

b The area of the lawn is 60 m².

Show that $x^2 + 5x - 60 = 0$.

c How long is the lawn? Give your answer to the nearest centimetre.

> **Hint** If you are asked to round your answer, the question will need to be solved using the quadratic formula.

4 **a** A square has sides of length $(x + 2)$ cm.

$(x + 2)$ cm

Find an expression for the area of the square in terms of *x*.

b A rectangle has sides of length $(2x + 4)$ cm and breadth $(x - 2)$ cm. The rectangle has the same area as the square in part **a**.

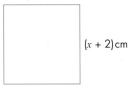

$(x - 2)$ cm

$(2x + 4)$ cm

Show that $x^2 - 4x - 12 = 0$.

c Hence find, algebraically, the dimensions of the square and rectangle.

20 Applying Pythagoras' theorem

Exercise 20A Applying Pythagoras' theorem

1 Is the triangle with sides of 9 cm, 40 cm and 41 cm a right-angled triangle?

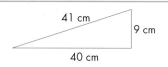

41 cm

9 cm

40 cm

2 Is this triangle right-angled? Justify your answer.

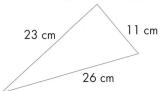

23 cm

11 cm

26 cm

3 Is this triangle right-angled? Justify your answer.

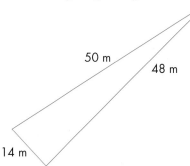

50 m

48 m

14 m

4 Jenny thinks her sandwich is in the shape of a right-angled triangle. She measures the sides. The sides measure 11 cm, 13 cm and 16 cm. Is Jenny correct? Justify your answer.

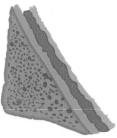

5 Tim checks to see if the large sail on his yacht is in the shape of a right-angled triangle. Its dimensions are 32 m, 34 m and 9 m.

Is it a right-angled triangle? Justify your answer.

6 A corridor is 5 m wide and turns through a right angle, as in the diagram.

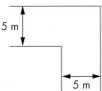

   a What is the longest pole that can be carried along the corridor horizontally?

   b If the corridor is 3 m high, what is the longest pole that can be carried along in any direction?

> Hint   Think about how to get the pole around the corner.

7 Calculate these lengths for the box shown.

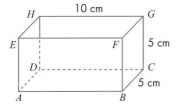

   a DG **b** HA **c** DB **d** AG

8 A cube has side 15 cm.

Calculate the distance between two vertically opposite corners.

9 Find the length of the space diagonal AB of a cuboid 9 cm by 9 cm by 40 cm.

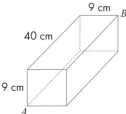

10 A small sculpture is made from four identical equilateral triangles of copper sheet stuck together to make a square-based pyramid.

The triangles have side 20 cm.

How high is the pyramid?

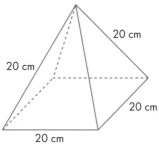

11 The diagram shows a square-based pyramid with base length 7 cm and sloping edges 12 cm. M is the midpoint of the side AB, X is the midpoint of the base, and E is directly above X.

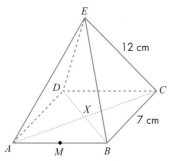

  a Calculate the length of the diagonal AC.

  b Calculate EX, the height of the pyramid.

  c Calculate the length EM.

  > Hint   Use triangle ABE.

12 The base of this pyramid, $ABCD$, is a square with side length 16 cm. The length of each sloping edge is 25 cm. The apex, V, is over the centre of the square base.

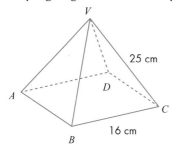

Calculate:

  a the height of the pyramid

  b the volume of the pyramid.

13 M is the midpoint of AB in this wedge.

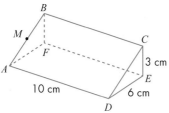

Calculate:

  a length CD **b** length DM.

14 A garage is 5 m long, 5 m wide and 2 m high.

Can a pole 7 m long be stored in it?

21 Applying the properties of shapes to determine an angle involving at least two steps

Example

A circle has centre O.

TK is a tangent to the circle.

Find the size of angle LHO.

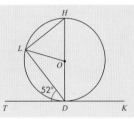

Strategy 1

Angle $LDO = 90 - 52 = 38°$

(Tangent and radius meet at 90°.)

Angle $OLD = 38°$

(Triangle OLD is isosceles as $OL = OD$ (radii).)

Angle $OLH = 52°$

(Triangle HLD is right-angled at L.)

Angle $LHO = 52°$

(Triangle LHO is isosceles as $OL = OH$ (radii).)

Strategy 2

Angle $LDO = 90 - 52 = 38°$

Angle $LOD = 180 - (2 \times 38) = 104°$

(Angles in a triangle add to 180°.)

Angle $LOH = 180 - 104 = 76°$

(Angles on a straight line add to 180°.)

Angle $LHO = \dfrac{180 - 76}{2} = 52°$

(Angles in a triangle add to 180° and triangle LHO is isosceles as $OL = OH$ (radii).)

> **Hint** Annotate the diagram with angles as you calculate them.
>
> For the final National 5 exam, working written in your diagram is accepted, but you must write the final answer with the diagram.

Exercise 21A Using angle properties of circles

 1 A circle has centre O.

CE is a tangent to the circle.

Calculate the size of angle BDE.

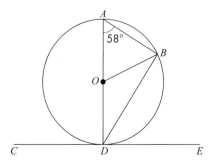

2 A circle has centre O.

HK is a tangent to the circle. EJ is a diameter.

Calculate the size of angle GEF.

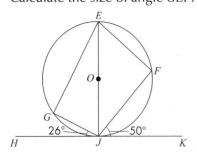

3 A circle has centre O.

HJ is a tangent to the circle. PR is a diameter.

Find the size of angle KPQ.

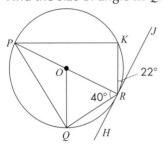

4 A circle has centre O.

SP and MP are tangents to the circle.

Find the size of angle SPM.

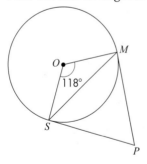

5 A circle has centre O.

TL and KL are tangents to the circle.

Find the size of angle KLT.

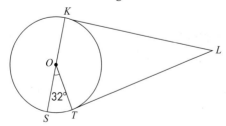

6 A circle has centre O.

HB and TB are tangents to the circle.

Find the size of angle SOT.

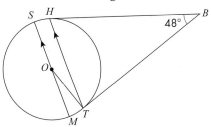

7 A circle has centre O.

WP and DC are parallel. DB is a tangent to the circle.

Find the size of angle DBC.

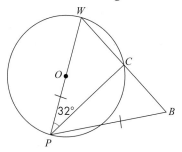

8 A circle has centre O.

$PW = PB$.

Find the size of angle PBC.

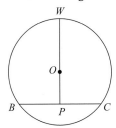

Exercise 21B Using Pythagoras' theorem

 1 A circle has centre O with radius $7\,$cm.

P is the midpoint of BC. $WP = 12.8\,$cm.

Find the length of BC. Give your answer to 2 decimal places (d.p.).

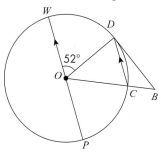

Hint	Find the length of OP by subtracting the radius from WP ($12.8\,$cm $- 7\,$cm).
	Draw the radius OB and find the length of BP using Pythagoras' theorem.
	Multiply your answer by 2.

2 A circle has centre *O* with radius 12 cm.

G is the midpoint of *NE*.

GK = 4 cm.

Find the length of *NE*. Give your answer to 1 d.p.

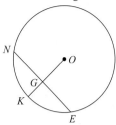

3 A circle has centre *O* and radius 11 cm.

T is the midpoint of *SP*.

SP = 15 cm.

Find the length of *w*. Give your answer to 3 significant figures (s.f.).

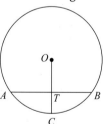

4 A circle has centre *O*.

T is the midpoint of *AB*.

AB = 24 cm and *OC* = 15 cm.

Find the length of *TC*.

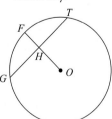

5 A circle has centre *O* with radius *r* cm.

GT = 16 cm and *FH* = 3 cm.

a Find an expression for *OH* in terms of *r*.

b Hence find the radius of the circle. Give your answer to an appropriate degree of accuracy.

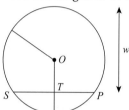

6 A tunnel has a cross-section formed by part of a circle with radius 8 m.

Find the height of the tunnel, h. Give your answer to 1 d.p.

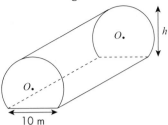

7 An oil container is in the shape of a cylinder with radius 1.4 m.

Calculate AB, the maximum width of oil in the tank. Give your answer to 2 d.p.

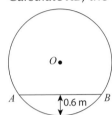

Exercise 21C Angles in polygons

1 Find the size of the angle marked with a letter in each quadrilateral.

a

b

c

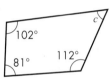

d

e

f

2 State whether each set of angles form the four interior angles of a quadrilateral. Give reasons for your answers.

a 125°, 65°, 70° and 90°   **b** 100°, 60°, 70° and 130°

c 85°, 95°, 85° and 95°   **d** 120°, 120°, 70° and 60°

e 112°, 68°, 32° and 138°   **f** 151°, 102°, 73° and 34°

3   **a** Give a reason why the sum of the interior angles of any pentagon is 540°.

    b Find the size of angle x in this pentagon.

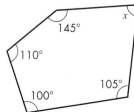

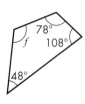

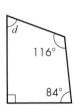

4 Calculate the size of the angle marked with a letter in each polygon.

a

b

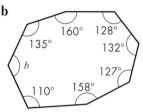

5 A regular octagon has eight sides.

a How many triangles can this shape be split into by drawing diagonals from one vertex?

b What is the sum of the interior angles of the octagon?

c What is the size of one interior angle?

6 A regular shape has 12 sides.

a How many triangles can this shape be split into by drawing diagonals from one vertex?

b What is the sum of the interior angles of the 12-sided shape?

c What is the size of one interior angle?

7 A regular shape has 30 sides.

a How many triangles can this shape be split into by drawing diagonals from one vertex?

b What is the sum of the interior angles of the 30-sided shape?

c What is the size of one interior angle?

8 For each regular polygon below, find the size of the interior angle x and the exterior angle y.

a   **b**   **c**

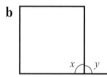 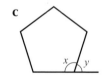

d   **e**

9 Find the number of sides of the regular polygon with an exterior angle of:

a 20°   **b** 30°   **c** 18°   **d** 4°

10 Find the number of sides of the regular polygon with an interior angle of:

a 135°   **b** 165°   **c** 170°   **d** 156°

11 *ABCDE* is a regular pentagon.

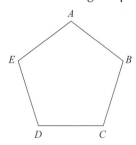

Work out the size of angle *ADE*. Justify your answer.

21 Applying the properties of shapes to determine an angle involving at least two steps

22 Using similarity

Exercise 22A Similar shapes

1 Each pair of shapes is similar. Find the lengths of the sides marked x.

a
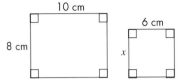

10 cm

6 cm

8 cm

x

b
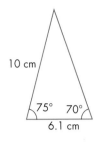

10 cm

75° 70°

6.1 cm

8 cm

75° 70°

x

2 Zahid's picture measures 12 cm by 8 cm. He wants to make a frame for his picture from wood that is 10 cm wide.

What length of wood does Zahid need to make the frame? Assume that the picture and the framed picture are similar shapes.

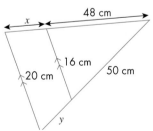

3 Calculate the length of x.

a
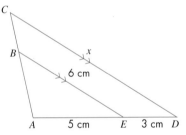

C

B

x

6 cm

A   5 cm   E   3 cm D

b
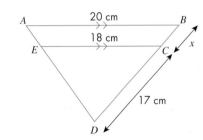

A   20 cm   B

18 cm

E   C x

17 cm

D

> **Hint**   Separate the triangles to make it easier.

4 Calculate the lengths of the sides marked x and y in these diagrams.

a
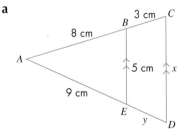

3 cm   C

B

8 cm

A

5 cm x

9 cm

E y D

b

48 cm

x

16 cm

20 cm   50 cm

y

5 Jamie is making this metal frame for a garden slide.

In the diagram, triangle *ABC* is similar to triangle *ADE*.

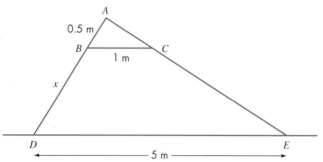

What length of metal does Jamie need to make *BD*, marked *x* on the diagram?

6 Suzie says that triangle *ABC* is similar to triangle *CDE*.

Is she correct?

Show working to explain your answer.

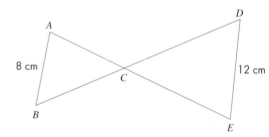

7 Triangle *ABC* is similar to triangle *CDE*.

The length of *BD* is 25 cm.

Work out the lengths of *BC* and *CD*.

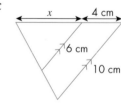

8 Calculate the lengths marked *x* in the diagrams below.

a

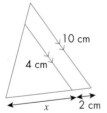

b

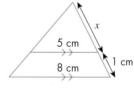

c

Example

A shop sells bottles of tomato ketchup which are mathematically similar.

The smaller bottle has a height of 15 cm and a volume of 189 ml.

The larger bottle has a height of 25 cm.

Find the volume of the larger bottle.

Linear scale factor $= \dfrac{25}{15}$

The linear scale factor (SF) is the fraction or ratio of two corresponding values. There is no need to simplify this unless you are carrying out a non-calculator question.

Volume scale factor $= \left(\dfrac{25}{15}\right)^3$

Volume SF = linear SF cubed.

Volume of larger bottle $= \left(\dfrac{25}{15}\right)^3 \times 189 = 875\,\text{ml}.$

Exercise 22B Areas and volumes of similar shapes

1 Don marks out a flower bed with an area of $20\,\text{cm}^2$. He then decides that he wants to use more space.

What would be the area of a similar flower bed with lengths that are four times the corresponding lengths of the first shape?

2 A brick has a volume of $400\,\text{cm}^3$.

What would be the volume of a similar brick with lengths that are:

a three times the corresponding lengths of the first brick

b five times the corresponding lengths of the first brick?

3 A tin of paint, 12 cm high, holds 4 litres of paint.

Show that a similar tin 36 cm high would hold 108 litres of paint.

4 A model statue is 15 cm high and has a volume of $450\,\text{cm}^3$. The real statue is 4.5 m high.

What is the volume of the real statue? Give your answer in m^3.

5 Tim has a large tin full of paint and wants to transfer the paint into a number of smaller tins like the one in the diagram.

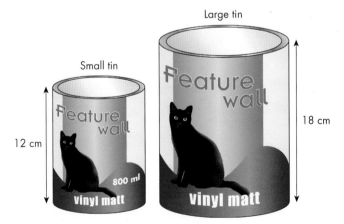

How many small tins can he fill from one large tin?

6 The mast on Erin's scale model yacht is 40 cm high. She dreams of owning the real yacht, which has a mast that is 4 m high.

a The sail on her model yacht has area $600\,\text{cm}^2$.

What would be the area of the sail on the real yacht? Give your answer in m^2.

b The volume of the hull on the real yacht is $20\,\text{m}^3$.

What is volume of the hull on her model yacht? Give your answer in cm^3.

7 The side length of a cube increases by 10%.

   a What is the percentage increase in the total surface area of the cube?

   b What is the percentage increase in the volume of the cube?

8 Standard and large gift boxes are mathematically similar.

The length of a large gift box is 15 cm.

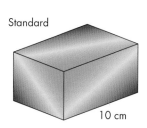

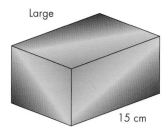

Standard

Large

10 cm

15 cm

The volume of the standard box is 240 cm³.

Which of the following is the correct volume of the large box?

   i 360 cm³   **ii** 540 cm³   **iii** 720 cm³   **iv** 810 cm³

9 A firm makes three sizes of similar-shaped bottles. Their volumes are:

Small: 330 cm³   Medium: 1000 cm³   Large: 2000 cm³

   a The medium bottle is 20 cm high.

   Calculate the heights of the other two bottles.

   b The firm designs a label for the large bottle and wants the labels on the other two bottles to be similar.

   If the area of the label on the large bottle is 100 cm², work out the area of the labels on the other two bottles.

10 Marie has two similar photographs.

28 cm

x

The areas of the photographs are 200 cm² and 600 cm².

Calculate the length of the side marked x marked on the smaller photograph.

11 These two bottles of cola are similar in shape.

If the height of one of the bottles is 20 cm, calculate the two possible heights of the other bottle.

550 ml   850 ml

23 Working with the graphs of trigonometric functions

Exercise 23A Graphs of trigonometric functions

1 Sketch the graphs of the following for $0 \leqslant x \leqslant 360$.

    a $y = 3\sin x°$
    b $y = 4\cos x°$
    c $y = \frac{2}{3}\sin x°$
    d $y = -2\cos x°$

2 State the equations of the following graphs.

    a
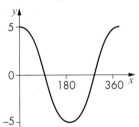

    b
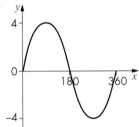

    c
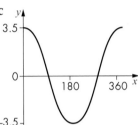

    d
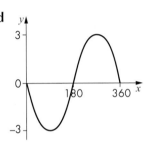

3 Sketch the graphs of the following for $0 \leqslant x \leqslant 360$.

    a $y = \sin 2x°$
    b $y = \cos 3x°$
    c $y = \cos \frac{1}{2}x°$
    d $y = \tan 2x°$

4 State the equations of the following graphs.

    a
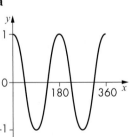

    b
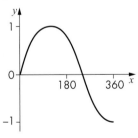

    c
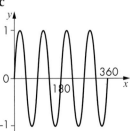

    d
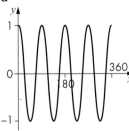

5 Sketch the graphs of the following for $0 \leqslant x \leqslant 360$.

    a $y = 4\sin 2x°$ **b** $y = 5\cos 4x°$ **c** $y = 1.5\sin 3x°$ **d** $y = -4\cos \frac{1}{2}x°$

6 State the equations of the following graphs.

a

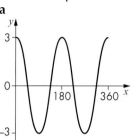

b

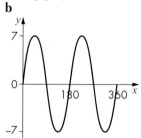

c

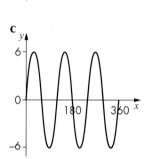

d

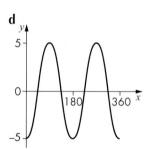

Example

The graph of $y = a\sin(x + b)°$, $0 \leqslant x \leqslant 360$, is shown.

a Write down the values of a and b.

b State the period of the graph.

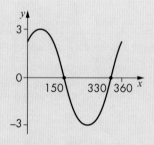

a $a = 3$

    $b = 30°$

b Period $= 360°$

> a represents the amplitude of the graph (the distance between the middle of the graph and the maximum value of the graph).

> Sketch the graph $y = 3\sin x$ and compare the roots with the given graph. Roots of $y = 3\sin x$ are $0°$, $180°$ and $360°$, so here the graph of $y = 3\sin x$ has been translated $30°$ to the left. A translation to the right would give a negative value for b.

> The length of one cycle of the graph. Graphs of the form $y = \sin bx$ or $y = \cos bx$ have a period of $\frac{360°}{b}$

Exercise 23B Translations of graphs of trigonometric functions

1 Sketch the graphs of the following for $0 \leqslant x \leqslant 360$.

    a $y = 2\sin x° + 3$ **b** $y = \cos 2x° - 3$ **c** $y = 4\sin \frac{1}{2}x° + 2$ **d** $y = 1 + 2\cos 3x°$

2 State the equations of the following graphs.

a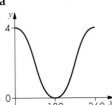

b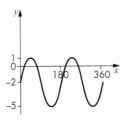

c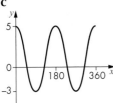

d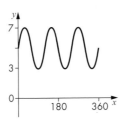

3 For each graph, **a–d**, in Question 2:

  i state the amplitude of the graph

  ii state the period of the graph.

4 Sketch the graphs of the following for $0 \leqslant x \leqslant 360$.

  a $y = \sin(x - 30)°$ **b** $y = 2\cos(x + 40)°$

  c $y = \cos(x - 20)° + 2$ **d** $y = 3\sin(x + 20)° - 4$

5 **a** The graph of $y = a\sin(x + b)°$, $0 \leqslant x \leqslant 360$, is shown.
    Write down the values of a and b.

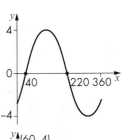

  b The graph of $y = \sin(x + b)° + c$, $0 \leqslant x \leqslant 360$, is shown.
    Write down the values of b and c.

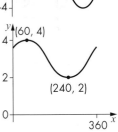

  c The graph of $y = a\cos(x + b)°$, $0 \leqslant x \leqslant 360$, is shown.
    Write down the values of a and b.

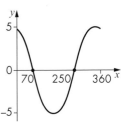

  d The graph of $y = \cos(x + b)° + c$, $0 \leqslant x \leqslant 360$, is shown.
    Write down the values of b and c.

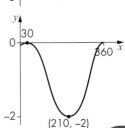

24 Working with trigonometric relationship in degrees

Example

Solve $3\sin x° + 1 = 2$, $0 \leqslant x \leqslant 360$. Give your answers to 1 decimal place (1 d.p.).

$3\sin x° = 1$ — Rearrange to solve for $\sin x°$.

$\sin x° = \frac{1}{3}$

Method 1: Sketch graph

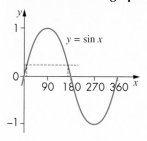

Method 2: ASTC diagram

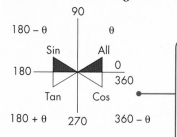

Use a sketch graph or ASTC diagram to find which quadrants the solutions lie in. $\sin x°$ is positive $\left(\frac{1}{3}\right)$ so both diagrams show that the two solutions of x lie in the 1st and 2nd quadrants.

1st quadrant angle

$x = \sin^{-1}\left(\frac{1}{3}\right) = 19.5°$ (1 d.p.)

2nd quadrant angle

$x = 180 - 19.5 = 160.5°$

Solutions: $x = 19.5°, 160.5°$ (1 d.p.)

Example

Solve $5\tan x° + 9 = 3$, $0 \leqslant x \leqslant 360$. Give your answers to 1 decimal place (1 d.p.).

$5\tan x° = -6$ — Rearrange.

$\tan x° = -\frac{6}{5}$

$\tan x° = \frac{6}{5}$

1st quadrant angle (θ)

$\tan \theta = \frac{6}{5}$

$\theta = \tan^{-1}\left(\frac{6}{5}\right) = 50.2°$

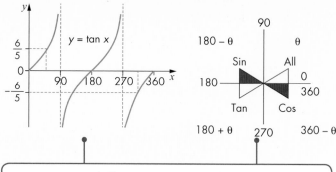

$\text{Tan}\,x°$ is negative $\left(-\frac{6}{5}\right)$ so the diagrams show that the two solutions of x lie in the 2nd and 4th quadrants.

First find the acute angle, θ, from the 1st quadrant, so you can use it to work out the solutions of x in the other quadrants.

2nd quadrant angle

$x = 180 - \theta = 180 - 50.2 = 129.8°$

4th quadrant angle

$x = 360 - \theta = 360 - 50.2 = 309.8°$

Solutions: $x = 129.8°, 309.8°$ (1 d.p.)

Exercise 24A Solving trigonometric equations

1 Solve the following equations, $0 \leqslant x \leqslant 360$, giving your answers to 1 d.p.

a $\sin x° = 0.452$ **b** $\cos x° = 0.75$ **c** $\tan x° = 2.3$ **d** $\sin x° = 0.62$

e $\cos x° = 0.912$ **f** $\sin x° = \frac{2}{5}$ **g** $\tan x° = \frac{5}{3}$ **h** $\cos x° = \frac{3}{7}$

2 Solve the following equations, $0 \leqslant x \leqslant 360$, giving your answers to 1 d.p.

a $\sin x° = -0.824$ **b** $\cos x° = -0.38$ **c** $\tan x° = -1.6$ **d** $\sin x° = -0.27$

e $\cos x° = -0.853$ **f** $\sin x° = -\frac{2}{3}$ **g** $\tan x° = -\frac{9}{4}$ **h** $\cos x° = -\frac{3}{8}$

3 Solve the following equations, $0 \leqslant x \leqslant 360$, giving your answers to 1 d.p.

a $6\sin x° - 5 = 0$ **b** $3\cos x° - 1 = 1$ **c** $2\tan x° - 6 = 2$ **d** $2\sin x° + 4 = 5$

e $7\cos x° - 5 = -3$ **f** $4\sin x° - 1 = 2$ **g** $6\tan x° - 9 = 0$ **h** $11\cos x° + 10 = 12$

4 Solve the following equations, $0 \leqslant x \leqslant 360$, giving your answers to 1 d.p.

a $2\sin x° + 3 = 2$ **b** $4\cos x° - 2 = -3$ **c** $\tan x° + 7 = 5$ **d** $7\sin x° + 3 = -2$

e $9\cos x° + 5 = 3$ **f** $8\sin x° + 8 = 5$ **g** $3\tan x° + 5 = 1$ **h** $10\cos x° - 4 = -7$

5 Below is the graph of $y = 3\sin x° + 2$, $0 \leqslant x \leqslant 360$.

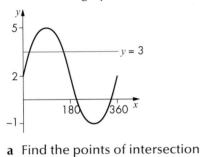

a Find the points of intersection with the line $y = 3$.

> **Hint**   Substitute $y = 3$ into the equation to solve.

b Find the coordinates of the points of intersection with the x-axis.

> **Hint**   Substitute $y = 0$ into the equation to solve.

6 The height of one of the cars of a Ferris wheel at a fairground is given by the formula

$h = 8 + 6\cos t°$, $0 \leqslant t \leqslant 360$

where h is the height in metres of the Ferris wheel above the ground after t seconds.

a What is the maximum height the Ferris wheel reaches?

> **Hint**   Consider the shape of the graph of
> $h = 8 + 6\cos t°$.
> Look back at Chapter 23 for help.
> What is the maximum value it reaches?

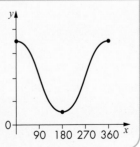

b When the Ferris wheel starts to turn, at what times will the car on the Ferris wheel reach a height of 6.8 metres in the first rotation?

7 The depth d, in metres, of water in a harbour was recorded from 12 noon. The depth was found to be given by the formula

$d = 7 + 4\sin^\circ t,\ 0 \leqslant t \leqslant 360$

where t is the number of minutes from the start of the recorded time.

a What is the depth of the water at 12 noon?

b Calculate the depth of water at 1 p.m.

c At which two times will the water have a depth of 5 m?

Example

Simplify $\dfrac{\sin^2 x}{1 - \sin^2 x}$ •————— | Use the trigonometric identities $\dfrac{\sin x}{\cos x} = \tan x$ or $\sin^2 x + \cos^2 x = 1$. |

$\dfrac{\sin^2 x}{1 - \sin^2 x} = \dfrac{\sin^2 x}{\cos^2 x}$ •————— | Use $\sin^2 x + \cos^2 x = 1$ and rearrange to substitute $\cos^2 x = 1 - \sin^2 x$. |

$= \left(\dfrac{\sin x}{\cos x}\right)^2$

$= \tan^2 x$ •————— | Substitute $\dfrac{\sin x}{\cos x} = \tan x$. |

Exercise 24B Trigonometric identities

 1 Simplify.

a $6\sin^3 x + 6\sin x \cos^2 x$

b $\tan x \times \dfrac{\cos^2 x}{\sin x}$

c $\dfrac{1 - \cos^2 x}{\cos^2 x}$

d $3 - 7\sin^2 x - 7\cos^2 x$

 2 Show that:

a $2 + 3\cos^2 x = 5 - 3\sin^2 x$

b $\dfrac{5\sin^3 x + 5\sin x \cos^2 x}{\cos x} = 5\tan x$

c $\dfrac{\tan x}{\sin x} = \dfrac{1}{\cos x}$

d $4\sin^2 x + 7\cos^2 x = 7 - 3\sin^2 x$

25 Calculating the area of a triangle using trigonometry

Exercise 25A Calculating the area of a triangle using trigonometry

1 Calculate the area of each triangle.

> **Hint** Use the formula for area, $A = \frac{1}{2}ab\sin C$.

a

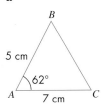

b

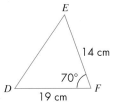

c

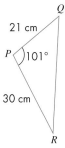

2 Calculate the area of each triangle.

a

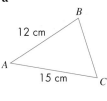

$\sin A = \frac{1}{2}$

b

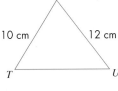

$\sin S = 0.8$

c

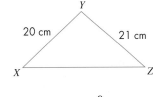

$\sin Y = \frac{9}{10}$

3 Work out the area of these triangles.

a Triangle ABC where $BC = 8\,\text{cm}$, $AC = 10\,\text{cm}$ and angle $ACB = 69°$.

b Triangle PQR where angle $QPR = 112°$, $PR = 3\,\text{cm}$ and $PQ = 7\,\text{cm}$.

4 Calculate the area of a regular hexagon of side length 7 cm.

> **Hint** Sketch a regular hexagon and split it into 6 triangles.

5 A regular octagon is shown.

The four diagonals of the octagon which are shown as dotted lines each have length 18 cm.

Find the area of the octagon.

6 **a** The area of triangle *ABC* is 71.8 cm².

Find the length of *AC*.

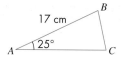

b The area of triangle *GHJ* is 109.9 cm².

Find the length of *HJ*.

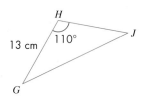

c The area of triangle *ABC* is 27 cm².

If *BC* = 12 cm and angle *BCA* = 98°, calculate *AC*.

Hint   Sketch a diagram to help you.

7 **a** The area of triangle *TKP* is 48 cm². *TK* = 18 cm and $\sin T = \frac{1}{3}$.

Find the length of *TP*.

b The area of triangle *BHT* is 108 cm². *HT* = 12 cm and $\sin H = \frac{3}{4}$.

Find the length of *BH*.

8 **a** The area of acute-angled triangle *FCT* is 30 cm². *FC* = 7 cm and *CT* = 9 cm.

Find the size of angle *FCT*.

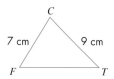

b The area of triangle *AHN* is 645 cm². *HA* = 40 cm and *HN* = 90 cm.

Find the size of angle *AHN*.

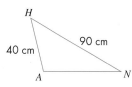

c The area of triangle *KLM* is 287 cm². *KL* = 25 cm and *LM* = 28 cm.

Find the size of obtuse angle *KLM*.

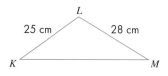

Hint   Remember that an obtuse angle is between 90° and 180°.

9 The area of acute-angled triangle *LMN* is 85 cm². *LM* = 10 cm and *MN* = 25 cm.

Calculate the size of angle *LMN*.

26 Using the sine and cosine rules to find a side or angle

Example

Calculate the length of the side marked x.

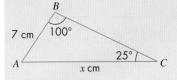

Label the diagram.

$B = 100°$, $C = 25°$, $b = x\,\text{cm}$, $c = 7\,\text{cm}$ ● Write down the values of B, C, b and c.

$$\dfrac{a}{\sin A} = \dfrac{b^{\checkmark}}{\sin B_{\checkmark}} = \dfrac{c^{\checkmark}}{\sin C_{\checkmark}}$$

Write down the sine rule and tick sides and angles with known information from the diagram.

$$\dfrac{x}{\sin 100°} = \dfrac{7}{\sin 25°}$$ ● Substitute known values into the formula.

$$x = \dfrac{7\sin 100°}{\sin 25°}$$ ● Rearrange to solve for x.

$x = 16.3\,\text{cm}$

Exercise 26A Using the sine rule

1 Calculate the length of the side or the size of the angle marked x in each triangle.

a
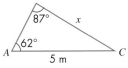

b
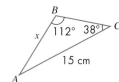

c
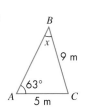

d
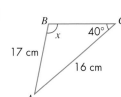

2
a In triangle ABC, $BC = 12$ cm, $\sin A = 0.6$ and $\sin B = 0.8$.
Find the length of AC.

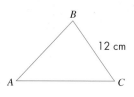

b In triangle TBC, $TC = 8$ cm, $\sin T = \frac{3}{4}$ and $\sin B = \frac{2}{3}$.
Find the length of BC.

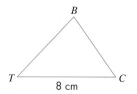

3 In a triangle ABC, angle A is $40°$, side AB is 10 cm and side BC is 7 cm.

> **Hint** Sketch and label a diagram to help you.

a If the triangle is acute-angled at C, calculate the size of angle ACB.

b If the triangle is obtuse-angled at C, calculate the size of angle ACB.

4 Triangle ABC has an obtuse angle at A.

Calculate the size of angle BAC.

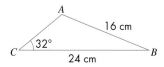

5 To calculate the length of a submarine, Mervyn stood at the top of a vertical cliff 60 m high and made some measurements (see diagram).

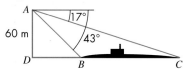

a Calculate the size of angle DAB.

b Calculate the length AB.

c Use the sine rule to work out the length of the submarine.

6 Use the information on this sketch to help the land surveyor calculate the width, w, of the river.

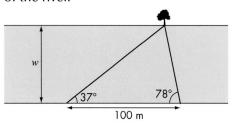

7 A surveyor needs to find the height of a chimney. She measures the angle of elevation as $28°$. She then walks 30 m towards the chimney and measures the angle of elevation from this point as $37°$.

What is the height of the chimney?

8 In a quadrilateral $ABCD$, $DC = 3\,\text{cm}$, $BD = 8\,\text{cm}$, angle $BAD = 43°$, angle $ABD = 52°$ and angle $BDC = 72°$.

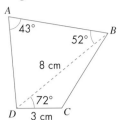

a Find the length of AD.

> **Hint** Use the sine rule:
> $$\frac{a}{\sin A} = \frac{b}{\sin B} = \frac{c}{\sin C}$$

b Hence find the area of the quadrilateral.

> **Hint** Find the area of each triangle then add together. See Chapter 25 for more help.

Exercise 26B Using the cosine rule

1 Calculate the length of the side marked x in each triangle.

a

7 m, 9 m, $72°$, x — triangle ABC

b

B — 25 cm — C, 40 cm, $110°$, x — triangle A

> **Hint** Use the cosine rule:
> $$a^2 = b^2 + c^2 - 2bc\cos A$$

2 Calculate the size of angle x in each triangle.

a

6 m, 11 m, 8 m, x — triangle ABC

b

B — 18 cm — C, 20 cm, 32 cm, x — triangle A

> **Hint** Use the rearranged cosine rule:
> $$\cos A = \frac{b^2 + c^2 - a^2}{2bc}$$

3 In the triangle given, show that $\cos B = -\frac{1}{5}$

4 cm, 5 cm, 7 cm — triangle ABC

4 The diagram shows a trapezium $ABCD$.
$AB = 6\,\text{cm}$, $AD = 8\,\text{cm}$, $CB = 12\,\text{cm}$ and angle $DAB = 115°$.

Calculate each of the following.

D — 8 cm — A, $115°$, 6 cm, C, 12 cm, B

a length DB
b angle DBA
c angle DBC
d length DC
e the area of the trapezium.

5 Harry is travelling on a road which goes directly from X to Y. This road is closed between A and B because of flooding, so Harry has to make a detour via C.

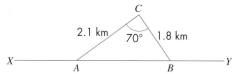

How much further does Harry have to travel as a result of this detour?

6 A quadrilateral $ABCD$ has $AD = 8\,cm$, $DC = 10\,cm$, $AB = 12\,cm$ and $BC = 15\,cm$. Angle $ADC = 112°$.

Calculate the size of angle ABC.

7 The three sides of a triangle are given as $3a$, $5a$ and $6a$.

Calculate the size of the smallest angle in the triangle.

8 Calculate the size of the smallest angle in triangle XYZ.

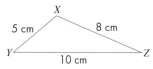

9 $ABCD$ is a quadrilateral.

Calculate the perimeter of $ABCD$.

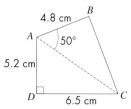

> **Hint** Find the length of AC first.

10 A sign writer wants to paint a board in the shape of a triangle.

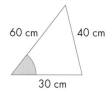

a Calculate the size of the shaded angle.

b Hence find the area of the board.

c A 50 ml pot of paint covers an area of approximately $0.5\,m^2$. The sign writer has only one pot in the required colour.

Will she have enough paint for a base coat of the sign? Give a reason for your answer.

Exercise 26C Choosing the correct formula

Sine rule	$\dfrac{a}{\sin A} = \dfrac{b}{\sin B} = \dfrac{c}{\sin C}$	to find a side when given two angles and a side *or* to find an angle when given two sides and an angle
Cosine rule	$a^2 = b^2 + c^2 - 2bc\cos A$	to find a side when given two sides and an angle
Cosine rule	$\cos A = \dfrac{b^2 + c^2 - a^2}{2bc}$	to find an angle when given three sides

1 Calculate the value of x in each triangle.

a
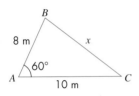

b
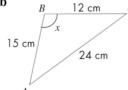

c
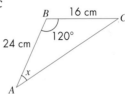

d
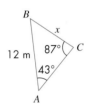

e
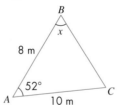

f
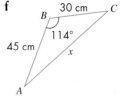

2 The hands of a clock have lengths 10 cm and 7 cm.

Work out the distance between the tips of the hands at 5 o'clock.

3 The diagram shows a sketch of quadrilateral $ABCD$.

Calculate:

a angle ABC **b** length AC.

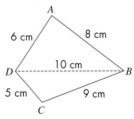

4 In triangle ABC, $AC = 7.6$ cm, angle $BAC = 35°$ and angle $ACB = 65°$.

Calculate the length AB.

5 Show that triangle ABC does not have an obtuse angle.

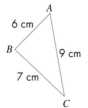

27 Using bearings with trigonometry

Example

In the diagram, $AB = 25\,\text{km}$, $BC = 18\,\text{km}$, B is on a bearing of $60°$ from A and C is on a bearing of $116°$ from B.

a Calculate the size of angle ABC.

b Calculate the length of AC.

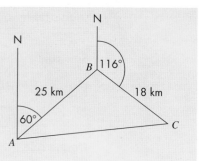

a

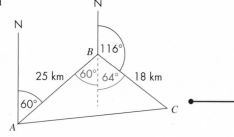

> Annotate the diagram. Extend the north lines and look for alternate angles and angles on a straight line, labelling any angles you have found. Identify the measurements you need to form the required angle.

Angle $ABC = 60 + 64 = 124°$

b

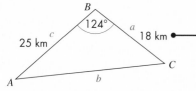

> Redraw the triangle showing only the necessary information.

$B = 124°$, $a = 18\,\text{km}$, $b = AC$, $c = 25\,\text{km}$

$b^2 = c^2 + a^2 - 2ac\cos B$

> Write down the information and state the version of the formula you will use.

$AC^2 = 25^2 + 18^2 - 2 \times 25 \times 18 \times \cos 124°$

> Substitute the values into the formula.

$AC^2 = 1452.27$

$AC = \sqrt{1452.27} = 38.1\,\text{km}$

Exercise 27A Using bearings with trigonometry

1 In the diagram, D is directly north of E.
What is the bearing of F from D?

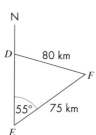

2 In the diagram, H is directly north of T.

The bearing of B from T is 296°.

Find the bearing of B from H.

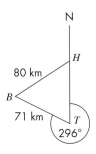

3 A boat leaves port K on a bearing of 108° and a yacht leaves port M on a bearing of 47°. They meet at point C.

How far has the yacht travelled?

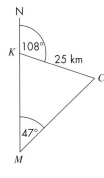

4 A man on a camel travels 14 km across the desert from village S to village D on a bearing of 76°. He then travels a further 9 km on a bearing of 112° to village G. He then returns directly back to village S.

a Calculate the size of angle SDG.

b Calculate the total distance travelled by the man and the camel.

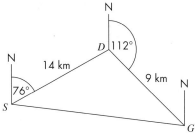

5 Three campsites are set up by groups of pupils participating in an expedition.

a Calculate the size of angle CPL.

b Hence find the bearing of L from P (the shaded angle).

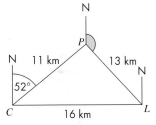

6 The diagram shows ship S and two lighthouses, A and B.

A is due west of B and the two lighthouses are 15 km apart.

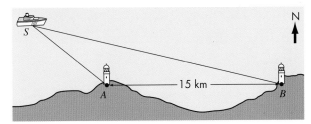

The bearing of the ship from lighthouse A is 330° and the bearing of the ship from lighthouse B is 290°.

How far is the ship from lighthouse B?

7 Port B is 20 km northeast of Port A. There is a lighthouse, L, 5 km from Port B on a bearing of 260°.

Calculate:

a the distance AL **b** the bearing of L from A to the nearest degree.

28 Working with 2D vectors

Example

Write each vector in component form.

a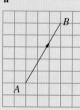

b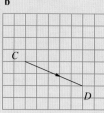

c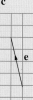

a $\overrightarrow{AB} = \begin{pmatrix} 3 \\ 5 \end{pmatrix}$ •————— Start from A, 3 units to right (positive) and 5 units up (positive).

b $\overrightarrow{CD} = \begin{pmatrix} 5 \\ -2 \end{pmatrix}$ •————— Start from C, 5 units to right (positive) and 2 units down (negative).

c $e = \begin{pmatrix} -1 \\ 4 \end{pmatrix}$ •————— Follow direction of arrow, 1 unit to left (negative) and 4 units up (positive).

Exercise 28A 2D vectors

1 Draw lines to represent these vectors. (Remember to include the arrow.)

$\mathbf{a} = \begin{pmatrix} 3 \\ 4 \end{pmatrix}$ $\qquad$ $\mathbf{b} = \begin{pmatrix} -3 \\ 4 \end{pmatrix}$ $\qquad$ $\mathbf{c} = \begin{pmatrix} 3 \\ -4 \end{pmatrix}$ $\qquad$ $\mathbf{d} = \begin{pmatrix} -3 \\ -4 \end{pmatrix}$

2 Write each vector in component form.

a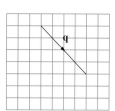

b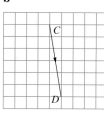

c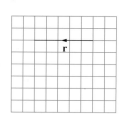

d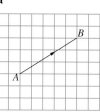

e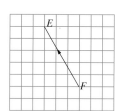

f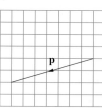

3 Draw the resultant vector of the following.

a **a** + **b**

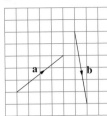

b **c** + **d**

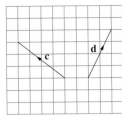

c **e** + **f**

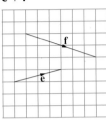

d **g** + **h** + **i**

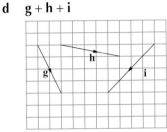

4 Draw the resultant vector of the following.

a **a** − **b**

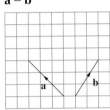

b **c** − **d**

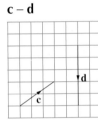

c **e** − **f**

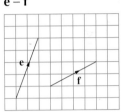

5 Draw the resultant vector of the following.

a **a** + 2**b**

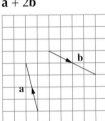

b 2**c** + 2**d**

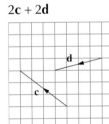

c 3**e** + 2**f**

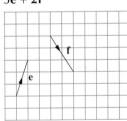

Exercise 28B Using vectors to solve problems

1 *OACB* is a trapezium.

$\vec{OA} = \mathbf{a}$, $\vec{OB} = \mathbf{b}$ and $\vec{BC} = 2\mathbf{a}$.

P and *Q* are the midpoints of $\vec{OB}$ and $\vec{AC}$.

a Express these vectors in terms of **a** and **b**.

   i $\vec{OP}$ **ii** $\vec{AQ}$ **iii** $\vec{PQ}$

b How can you tell that $\vec{PQ}$ is parallel to $\vec{OA}$?

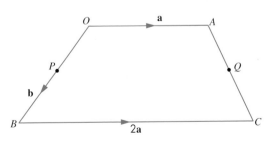

2 *ABCDEF* is a regular hexagon with centre *O*.
$\overrightarrow{OA} = \mathbf{a}$ and $\overrightarrow{OB} = \mathbf{b}$.

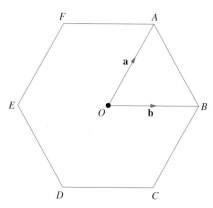

Write each of these vectors in terms of **a** and **b**.
Give your answers in their simplest form.

a $\overrightarrow{AB}$ **b** $\overrightarrow{AD}$ **c** $\overrightarrow{EC}$ **d** $\overrightarrow{FB}$

3 $\overrightarrow{OA} = \mathbf{a}$ and $\overrightarrow{OB} = \mathbf{b}$. Point C divides the line *AB* in the ratio 3 : 1.

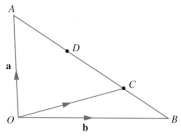

a Express $\overrightarrow{OC}$ in terms of **a** and **b**.

b If point *D* is the midpoint of *AC*, express $\overrightarrow{OD}$ in terms of **a** and **b**.

4 $\overrightarrow{OA} = 10\mathbf{q}$ and $\overrightarrow{OB} = 5\mathbf{p}$.
$\overrightarrow{AX} = 4\overrightarrow{XB}$.

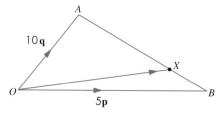

Write these vectors in terms of **p** and **q**.

a $\overrightarrow{AB}$ **b** $\overrightarrow{AX}$ **c** $\overrightarrow{OX}$

5 In the diagram, $\overrightarrow{OA}$ = **a** and $\overrightarrow{OB}$ = **b**.

Q is the midpoint of BC and the point P divides BA in the ratio 1 : 2.

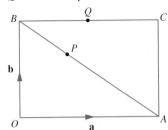

a Write these vectors in terms of **a** and **b**.

  i $\overrightarrow{BP}$ **ii** $\overrightarrow{OP}$ **iii** $\overrightarrow{OQ}$

b Describe the relationship between O, P and Q.

6 OAB is a triangle. P, Q and R are the midpoints of OB, AB and OA. $\overrightarrow{OR}$ = **a** and $\overrightarrow{OP}$ = **b**.

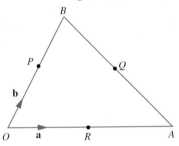

a Express these vectors in terms of **a** and **b**.

  i $\overrightarrow{RP}$ **ii** $\overrightarrow{AB}$

b Describe the relationship between $\overrightarrow{RP}$ and $\overrightarrow{AB}$.

7 $\overrightarrow{OC}$ = 12**q**, $\overrightarrow{OB}$ = 3**q** and $\overrightarrow{OA}$ = 3**p**.

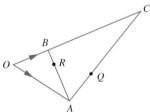

a Write these vectors in terms of **p** and **q**.

  i $\overrightarrow{AB}$ **ii** $\overrightarrow{AC}$

b Given that $AQ = \frac{1}{3}AC$, express $\overrightarrow{OQ}$ in terms of **p** and **q**.

c Given that $\overrightarrow{OR}$ = **p** + 2**q**, what can you say about the points O, R and Q?

29 Working with 3D coordinates

Exercise 29A Working with 3D coordinates

1. For each diagram, write down the coordinates of all the vertices shown.

a
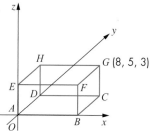

b
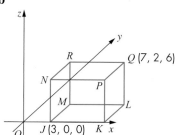

c
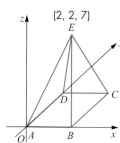

d
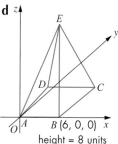
height = 8 units

e
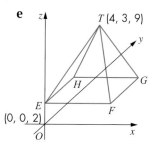

f
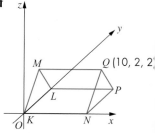

> **Hint** The shapes in parts **c** and **d** are square-based pyramids.
>
> In parts **b** and **e**, note that the cuboid does not touch the origin.
>
> In parts **c**, **d** and **e**, use the symmetry of the shape – the top vertex is in the middle.

2. a The diagram shows a cube placed on top of a cuboid, relative to the coordinate axes. *A* lies on the origin.

   Write down the coordinates of all the vertices shown.

   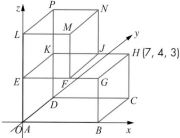

   b The diagram shows a triangular prism placed on top of a cuboid.

   A lies on the origin. Triangles *EHJ* and *FGK* are isosceles triangles (*EJ* = *HJ* and *FK* = *GK*). The height of the triangular prism is half the height of the cuboid.

   Write down the coordinates of all the vertices shown.

   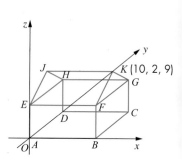

c The diagram below shows a square-based pyramid placed on top of a cuboid.
The cuboid and the square-based pyramid have the same height.
Write down the coordinates of vertices *C*, *D* and *E*.

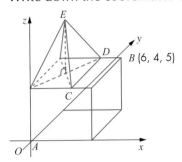

3 The following diagram shows a cuboid relative to the coordinate axes. *M* lies on the origin.

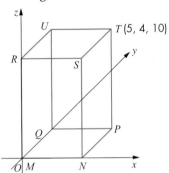

  a Find the coordinates of:

    i the midpoint of *RS*

    ii the midpoint of *SU*

    iii the midpoint of *PT*.

  b Find the length of *NU*. Give your answer correct to 3 significant figures.

> Hint   Use Pythagoras' theorem in three dimensions.

4 The diagram shows a square pyramid placed on top of a cube, relative to the coordinate axes. *A* lies on the origin.

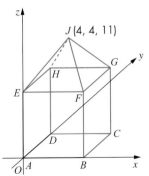

  a Write down the coordinates of all the vertices shown.

  b Find the length of *GJ*. Give your answer correct to 1 decimal place.

30 Using vector components

Example

Vectors **p** and **q** have components $\mathbf{p} = \begin{pmatrix} 3 \\ -2 \end{pmatrix}$ and $\mathbf{q} = \begin{pmatrix} -1 \\ 3 \end{pmatrix}$

Find the resultant vector $2\mathbf{p} + \mathbf{q}$.

$2\mathbf{p} = \begin{pmatrix} 6 \\ -4 \end{pmatrix}$ ⟵ Multiply each component of **p** by 2.

$2\mathbf{p} + \mathbf{q} = \begin{pmatrix} 6 \\ -4 \end{pmatrix} + \begin{pmatrix} -1 \\ 3 \end{pmatrix} = \begin{pmatrix} 6 + (-1) \\ (-4) + 3 \end{pmatrix}$ ⟵ Add the components.

$= \begin{pmatrix} 5 \\ -1 \end{pmatrix}$

Exercise 30A Using vector components

 1 Vectors **a**, **b** and **c** have components $\mathbf{a} = \begin{pmatrix} 2 \\ 3 \end{pmatrix}$, $\mathbf{b} = \begin{pmatrix} -3 \\ 5 \end{pmatrix}$ and $\mathbf{c} = \begin{pmatrix} -2 \\ -6 \end{pmatrix}$

Find the following resultant vectors. Give your answers in component form.

   a $\mathbf{a} + 2\mathbf{b}$ **b** $2\mathbf{b} + 3\mathbf{c}$

   c $3\mathbf{a} - 2\mathbf{c}$ **d** $2\mathbf{a} + \frac{1}{2}\mathbf{c}$

   e $\mathbf{a} + \mathbf{b} + \mathbf{c}$ **f** $4\mathbf{a} - 2\mathbf{b} + \mathbf{c}$

   g $-2\mathbf{a} - 4\mathbf{b}$ **h** $-\mathbf{a} - 3\mathbf{b} - \mathbf{c}$

 2 Vectors **u**, **v** and **w** have components $\mathbf{u} = \begin{pmatrix} 4 \\ 2 \\ -1 \end{pmatrix}$, $\mathbf{v} = \begin{pmatrix} -2 \\ 7 \\ -9 \end{pmatrix}$ and $\mathbf{w} = \begin{pmatrix} -3 \\ -6 \\ 3 \end{pmatrix}$

Find the following resultant vectors. Give your answers in component form.

   a $\mathbf{u} + 3\mathbf{v}$ **b** $3\mathbf{u} + 2\mathbf{w}$

   c $4\mathbf{v} - 3\mathbf{w}$ **d** $3\mathbf{u} + \frac{1}{3}\mathbf{w}$

   e $\mathbf{u} + \mathbf{v} - \mathbf{w}$ **f** $5\mathbf{u} - 3\mathbf{v} + 2\mathbf{w}$

   g $-3\mathbf{v} - 4\mathbf{w}$ **h** $-\mathbf{v} - 3\mathbf{w} - \mathbf{u}$

3 Calculate the missing values.

  a $\begin{pmatrix} 3 \\ -2 \\ a \end{pmatrix} - \begin{pmatrix} 5 \\ b \\ -4 \end{pmatrix} = \begin{pmatrix} -2 \\ -6 \\ 3 \end{pmatrix}$ **b** $\begin{pmatrix} 5 \\ -3 \\ z \end{pmatrix} + \begin{pmatrix} x \\ -2 \\ -1 \end{pmatrix} = \begin{pmatrix} 3 \\ y \\ -3 \end{pmatrix}$ **c** $\begin{pmatrix} -3 \\ 2a \\ 2b \end{pmatrix} + \begin{pmatrix} -1 \\ 5b \\ 3a \end{pmatrix} = \begin{pmatrix} -4 \\ -1 \\ 4 \end{pmatrix}$

> **Hint**   In part **c**, set up a pair of simultaneous equations to solve for a and b.

Exercise 30B Calculating the magnitude of vectors

1 Find the magnitude of the following vectors. Give your answers correct to 2 decimal places (2 d.p.).

> **Hint** $|\mathbf{a}| = \sqrt{x^2 + y^2}$ where $\mathbf{a} = \begin{pmatrix} x \\ y \end{pmatrix}$
>
> $|\mathbf{a}|$ means the magnitude of **a**.

a $\mathbf{a} = \begin{pmatrix} 2 \\ 5 \end{pmatrix}$ **b** $\mathbf{b} = \begin{pmatrix} 3 \\ -1 \end{pmatrix}$ **c** $\mathbf{c} = \begin{pmatrix} -4 \\ 6 \end{pmatrix}$ **d** $\mathbf{d} = \begin{pmatrix} -9 \\ -8 \end{pmatrix}$

2 Find the magnitude of the following vectors. Give each answer as a surd in its simplest form.

a $\overrightarrow{AB} = \begin{pmatrix} 2 \\ 4 \end{pmatrix}$ **b** $\overrightarrow{CD} = \begin{pmatrix} 3 \\ -6 \end{pmatrix}$ **c** $\overrightarrow{EF} = \begin{pmatrix} -6 \\ 2 \end{pmatrix}$ **d** $\overrightarrow{GH} = \begin{pmatrix} -9 \\ -3 \end{pmatrix}$

3 Find the magnitude of the following vectors. Give your answers correct to 2 d.p.

> **Hint** $|\mathbf{a}| = \sqrt{x^2 + y^2 + z^2}$ where $\mathbf{a} = \begin{pmatrix} x \\ y \\ z \end{pmatrix}$

a $\mathbf{p} = \begin{pmatrix} 2 \\ 4 \\ -3 \end{pmatrix}$ **b** $\mathbf{q} = \begin{pmatrix} 7 \\ -9 \\ -2 \end{pmatrix}$ **c** $\mathbf{r} = \begin{pmatrix} 11 \\ -2 \\ -4 \end{pmatrix}$ **d** $\mathbf{s} = \begin{pmatrix} -13 \\ -5 \\ -6 \end{pmatrix}$

4 Find the magnitude of the following vectors. Give each answer as a surd in its simplest form.

a $\overrightarrow{JK} = \begin{pmatrix} -2 \\ 5 \\ -4 \end{pmatrix}$ **b** $\overrightarrow{LM} = \begin{pmatrix} -1 \\ -4 \\ 1 \end{pmatrix}$ **c** $\overrightarrow{NP} = \begin{pmatrix} 2 \\ -2 \\ -6 \end{pmatrix}$ **d** $\overrightarrow{RS} = \begin{pmatrix} -8 \\ 6 \\ -8 \end{pmatrix}$

5 Vectors **a**, **b** and **c** have components $\mathbf{a} = \begin{pmatrix} 3 \\ -4 \end{pmatrix}$, $\mathbf{b} = \begin{pmatrix} -2 \\ 7 \end{pmatrix}$ and $\mathbf{c} = \begin{pmatrix} -1 \\ -8 \end{pmatrix}$

Calculate the following, giving your answers correct to 2 d.p.

> **Hint** $|\mathbf{a} + \mathbf{b}|$ means calculate the magnitude of $\mathbf{a} + \mathbf{b}$. Find the components of $\mathbf{a} + \mathbf{b}$, then find the magnitude of this resultant vector.

a $|\mathbf{a} + \mathbf{b}|$ **b** $|\mathbf{b} + 2\mathbf{c}|$ **c** $|2\mathbf{a} - 3\mathbf{c}|$ **d** $|\mathbf{a} - 2\mathbf{b} + \mathbf{c}|$

6 Vectors **p**, **q** and **r** have components $\mathbf{p} = \begin{pmatrix} 2 \\ -3 \\ 1 \end{pmatrix}$, $\mathbf{q} = \begin{pmatrix} -9 \\ 2 \\ -2 \end{pmatrix}$ and $\mathbf{r} = \begin{pmatrix} -3 \\ -2 \\ -3 \end{pmatrix}$

Calculate the following, giving your answers correct to 2 d.p.

a $|\mathbf{p} + \mathbf{q}|$ **b** $|2\mathbf{p} + \mathbf{r}|$ **c** $|3\mathbf{p} - 2\mathbf{q}|$ **d** $|2\mathbf{p} - \mathbf{q} + 2\mathbf{r}|$

31 Working with percentages

Example

A town has a population of 4220. It is expected that the population of the town will increase by 7% each year over the next 4 years.

What is the population expected to be after 4 years? Give your answer to the nearest 10.

$100\% + 7\% = 107\% = 1.07$

$4220 \times 1.07^4 = 5531.559...$

≈ 5530 (rounded to nearest 10)

The population would be 5530.

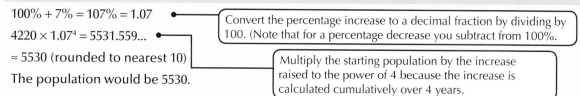

Convert the percentage increase to a decimal fraction by dividing by 100. (Note that for a percentage decrease you subtract from 100%.)

Multiply the starting population by the increase raised to the power of 4 because the increase is calculated cumulatively over 4 years.

> **Hint** You can also calculate the population by finding 7% of 4220, adding this amount to 4220, then repeating these steps for the specified number of years. However, the method demonstrated is more efficient.

Exercise 31A Percentage increase and decrease

1 A bank pays compound interest of 8% per annum (per year) on money invested in a savings account. Petra invests £2000 in this account.

How much will she have in her account after 3 years?

> **Hint** **Compound interest** means that interest is added every year, so each year the amount of interest being added increases.

2 A small plant increases its height by 10% each day in the second week of its growth. At the end of the first week, the plant was 5 cm high.

What is its height after a further:

a 1 day   **b** 2 days   **c** 4 days   **d** 1 week?

3 Billy put a gift of £250 into a special savings account that offered him 8% compound interest per annum if he promised to keep the money in the account for at least 2 years.

How much was in this account after:

a 2 years   **b** 3 years   **c** 5 years?

4 The penguin population of a small island was only 1500 in 2008, but increased steadily by about 15% each year.

Calculate the population in:

a 2009   **b** 2010   **c** 2012

5 I take out a loan for £14 499 to buy a car. The loan is for 5 years and has an interest rate of 6.9% per annum.

What will my total loan repayments be?

6 Veronika buys a mobile phone for £250. It depreciates at a rate of 11.3% each year.
Amelia buys a mobile phone for £320. It loses value at a rate of 8.9% each year.
Scarlett buys a phone for £650. It depreciates by 22% each year.

Whose phone is worth the most after 3 years?

7 I invest £3000 in a savings account that pays compound interest of 4% per annum.

How much will I have in the bank after 6 years if I do not make any withdrawals?

8 A chick increases its mass by 3% each day for the first 6 weeks of life. The mass of the chick was 85 g at birth.

What will its mass be after:

a 1 day   **b** 5 days   **c** 10 days?

9 The headteacher of a new school offered her staff an annual pay increase of 5% for every year they stayed with the school.

a Mr Speed started teaching at the school on a salary of £28 000.

What will his salary be after 3 years?

b Miss Tuck started teaching at the school on a salary of £14 500.

How many years will it be until she is earning a salary of over £20 000?

10 The population of a small village in Lanarkshire is projected to increase at a steady rate of 9% each year. In 2010 the population was 19 102.

a In which year is the population:

    i 22 695   **ii** 29 390   **iii** 45 221   **iv** 107 055?

b When the population of the village reaches 55 000, the council will need to build a new school.

In which year will this be?

11 My dad put all the money I was sent when I was born into a savings account. He invested £470 in an account that pays compound interest of 11.2% per annum.

How much will I have when I turn 18?

12 A sycamore tree is 40 cm tall. It grows at a rate of 8% per year. A conifer is 20 cm tall. It grows at a rate of 15% per year.

How many years does it take before the conifer is taller than the sycamore?

13 The population of a small town is 2000. It is falling by 10% each year. The population of a nearby village is 1500. It is rising by 10% each year.

After how many years will the population of the town be less than the population of the village?

14 Each week, a boy takes out 20% of the amount in his bank account to spend.

After how many weeks will the amount in his bank account have halved from the original amount?

Example

A laptop is reduced by 15% to £544 in a sale.

What was the original price of the laptop?

$100\% - 15\% = 85\%$ (The final cost is 85% of the original cost.)

$85\% = £544$

$1\% = \dfrac{544}{85} = £6.40$ (Find 1% of the original amount by dividing by 85.)

$100\% = 6.40 \times 100 = £640$ (Find 100% by multiplying by 100.)

The laptop originally cost £640.

Exercise 31B Reverse use of percentages

1 A carpenter increases his charges by 6% to £25.44 per hour.

How much did he charge before the increase?

2 13% more tickets were sold for Z Festival this year than last year. This year 58 082 tickets were sold.

How many tickets were sold last year?

3 Steve reduced his personal best marathon time by 4% to 3 hours and 36 minutes.

What was his previous personal best?

4 Every year the amount of rubbish sent to landfill increases by 7%. This year 25 265 696 tonnes of rubbish will be sent to landfill.

How much rubbish was sent to landfill last year?

5 Steve has reduced his calorie intake by 5% and now consumes 1995 calories per day.

What was his original calorie intake?

6 A pair of shoes are reduced by 30% to £105.

How much did they cost before the reduction?

7 I bought a sofa in the January sales with 25% discount. I paid £330.

What is the full price of the sofa?

8 Every year 20% of university students drop out of Course B at CleberClobs University. This year 44 students dropped out of Course B.

How many students started the course?

9 Paula spends £9 each week on CDs. This is 60% of her weekly allowance.

How much is Paula's weekly allowance?

10 Aaron is demoted at work and his salary decreases by 15% to £38 250.

How much was he earning before his demotion?

11 Yuni's weekly pay is increased by 4% to £187.20.

What was Yuni's pay before the increase?

12 Jon's salary is £23 980. This is 10% more than he earned 2 years ago. Last year his salary was 3% more than it was 2 years ago.

  a How much was his salary last year?

  b By what percentage did his salary increase last year?

13 A man's savings decreased by 10% in one year and then increased in the following year by 10%. He now has £1782.

How much did he have 2 years ago?

14 Twice as many people visit a shopping centre on Saturdays as on Fridays. The numbers visiting on both days increases by 50% in the week before Christmas.

How many more visit on this Saturday than on this Friday? Give your answer as a percentage.

32 Working with fractions

 All questions in this chapter should be completed without the use of a calculator unless specified.

All answers should be stated in their simplest form.

Example

Evaluate $3\frac{4}{5} + 1\frac{2}{3}$

$3\frac{4}{5} + 1\frac{2}{3} = 4 + \frac{4}{5} + \frac{2}{3}$ ●————(Add the whole numbers together and add the fractions.)

$= 4 + \frac{...}{15} + \frac{...}{15}$ ●————(Find the lowest common multiple of the denominators. 15 is the LCM of 5 and 3.)

$= 4 + \frac{12}{15} + \frac{10}{15}$ ●————(Find the numerators of the equivalent fractions.)

$= 4\frac{22}{15} = 4 + 1\frac{7}{15} = 5\frac{7}{15}$ ●————(Simplify the answer.)

Hint An alternative strategy is to convert both fractions to improper fractions, but this is not always the most efficient method.

Exercise 32A Adding and subtracting fractions

1 Evaluate the following.

a $\frac{1}{2} + \frac{1}{5}$ b $\frac{1}{2} + \frac{1}{3}$ c $\frac{1}{3} + \frac{1}{10}$ d $\frac{3}{8} + \frac{1}{3}$

e $\frac{3}{4} + \frac{1}{5}$ f $\frac{1}{3} + \frac{2}{5}$ g $\frac{3}{5} + \frac{3}{8}$ h $\frac{1}{2} + \frac{2}{5}$

2 Evaluate the following.

a $\frac{7}{8} - \frac{3}{4}$ b $\frac{4}{5} - \frac{1}{2}$ c $\frac{2}{3} - \frac{1}{5}$ d $\frac{3}{4} - \frac{2}{5}$

3 Evaluate the following.

a $\frac{3}{5} + \frac{7}{16} - \frac{1}{3}$ b $\frac{9}{24} + \frac{5}{18} - \frac{1}{10}$ c $\frac{1}{4} + \frac{7}{9} - \frac{5}{13}$

4 Evaluate the following.

a $5\frac{1}{4} + 7\frac{3}{5}$ b $8\frac{2}{3} + 1\frac{4}{9}$ c $6\frac{3}{4} + 2\frac{7}{10}$

d $9\frac{1}{8} + 3\frac{7}{25}$ e $7\frac{9}{20} + 3\frac{5}{16}$ f $8\frac{3}{8} + 1\frac{3}{16} + 2\frac{3}{4}$

g $6\frac{17}{20} - 5\frac{5}{12}$ h $2\frac{5}{8} - 1\frac{7}{24}$ i $3\frac{9}{32} - 1\frac{1}{12}$

j $4\frac{3}{5} + 5\frac{7}{16} - 8\frac{1}{3}$ k $1\frac{9}{24} + 1\frac{5}{18} - 1\frac{1}{10}$ l $5\frac{1}{4} + 2\frac{7}{9} - 6\frac{5}{13}$

5 **a** At a football club, half of the players are Scottish, a quarter are English and one-sixth are Italian. The rest are Irish.

What fraction of players at the club are Irish?

b Which of the following is the number of players at the club?

30   32   34   36

6 On a firm's coach trip, half the people were employees and two-fifths were partners of the employees. The rest were children.

What fraction were children?

7 Five-eighths of a crowd of 35 000 people were male.

How many females were in the crowd?

8 Look at this road sign.

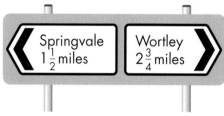

a What is the distance between Springvale and Wortley, using these roads?

b How much further is it to Wortley than to Springvale?

9 Pipes are sold in $\frac{1}{2}$m and $\frac{3}{4}$m lengths.

What is the least number of pipes that can be used to make a pipe exactly 2 m long? (Assume that you cannot cut pipes to size.) Show your working.

10 A class had the same number of boys and girls. Three girls leave the class and are replaced by three boys. Three-eighths of the class are now girls.

How many students are in the class?

Example

Evaluate $2\frac{2}{5} \times 1\frac{3}{4}$

$2\frac{2}{5} \times 1\frac{3}{4} = \frac{12}{5} \times \frac{7}{4}$ •————(Convert to improper fractions.)

$= \frac{{}^{3}\cancel{12}}{5} \times \frac{7}{\cancel{4}_{1}}$ •————(Simplify by cancelling. Multiply numerators and multiply denominators.)

$= \frac{21}{5} = 4\frac{1}{5}$ •————(Convert to a mixed number.)

Hint Division by a fraction is the same as multiplication by its reciprocal. For example $\div\frac{2}{3}$ changes to $\times\frac{3}{2}$

Exercise 32B Multiplying and dividing fractions

1 Evaluate the following.

a $\frac{1}{2} \times \frac{2}{3}$ b $\frac{3}{4} \times \frac{2}{5}$ c $\frac{3}{5} \times \frac{1}{2}$ d $\frac{3}{7} \times \frac{2}{3}$

e $\frac{2}{3} \times \frac{5}{6}$ f $\frac{1}{3} \times \frac{3}{5}$ g $\frac{2}{3} \times \frac{7}{10}$ h $\frac{3}{8} \times \frac{2}{5}$

i $\frac{4}{9} \times \frac{3}{8}$ j $\frac{4}{5} \times \frac{1}{2} \times \frac{3}{8}$ k $\frac{3}{4} \times \frac{7}{10} \times \frac{5}{6}$ l $\frac{2}{3} \times \frac{5}{6} \times \frac{9}{10}$

2 Evaluate the following.

a $\frac{1}{5} \div \frac{1}{3}$ b $\frac{3}{5} \div \frac{3}{8}$ c $\frac{4}{5} \div \frac{2}{3}$ d $\frac{4}{7} \div \frac{8}{9}$

e $\frac{1}{4} \div \frac{1}{3}$ f $\frac{4}{5} \div \frac{2}{10}$ g $\frac{1}{2} \div \frac{2}{4}$ h $\frac{3}{5} \div \frac{6}{10}$

3 Evaluate the following. Give your answer as a mixed number where possible.

a $1\frac{1}{3} \times 2\frac{1}{4}$ b $1\frac{3}{4} \times 1\frac{1}{3}$ c $2\frac{1}{2} \times \frac{4}{5}$ d $1\frac{2}{3} \times 1\frac{3}{10}$

e $3\frac{1}{4} \times 1\frac{3}{5}$ f $2\frac{2}{3} \times 1\frac{3}{4}$ g $3\frac{1}{2} \times 1\frac{1}{6}$ h $7\frac{1}{2} \times 1\frac{3}{5}$

i $2\frac{1}{6} \times 1\frac{2}{3}$ j $3\frac{2}{3} \times 3$ k $2\frac{2}{3} \times 3$ l $1\frac{1}{2} \times \frac{2}{3}$

4 Evaluate the following.

a $3\frac{1}{3} \div 2\frac{1}{2}$ b $4\frac{1}{3} \div 4\frac{1}{4}$ c $4\frac{4}{5} \div 2\frac{7}{10}$ d $4\frac{2}{5} \div 4\frac{3}{4}$

e $3\frac{3}{5} \div 2\frac{1}{2}$ f $3\frac{9}{10} \div 2\frac{2}{3}$ g $4\frac{1}{2} \div 4\frac{7}{10}$ h $4\frac{1}{5} \div 4\frac{4}{5}$

i $4\frac{1}{2} \div 4\frac{3}{4}$ j $3\frac{3}{5} \div 3\frac{3}{4}$ k $4 \div 1\frac{1}{2}$ l $5 \div 3\frac{2}{3}$

5 Kris walked three-quarters of the way along Carterknowle Road, which is 3 km long. How far did Kris walk?

6 Jean ate one-fifth of a cake, Les ate half of what was left, and Nick ate the rest. What fraction of the cake did Nick eat?

7 Which is smaller: $\frac{3}{4}$ of $5\frac{1}{3}$ or $\frac{2}{3}$ of $4\frac{2}{5}$? Give a reason for your answer.

8 I estimate that I need 60 litres of lemonade for a party. I buy 24 bottles, each containing $2\frac{3}{4}$ litres.
Have I bought enough lemonade for the party?

9 Sergio wants to put up a fence along one side of his garden, which is 20 m long. The fence comes in sections $1\frac{1}{3}$ m long.
How many sections will Sergio need to put the fence all the way down his garden?

10 An African bullfrog can jump a distance of $1\frac{1}{4}$ m in one hop.

How many hops would it take an African bullfrog to hop a distance of 100 m?

11 Three-fifths of all 14-year-olds in a school visit the dentist each year. One-third of those who do not visit the dentist have a problem with their teeth.

What fraction of all the 14-year-olds do not visit the dentist and have a problem with their teeth?

12 How many half-litre tins of paint can be poured into a 2.2 litre paint tray without spilling?

> **Hint** Change 2.2 into a fraction. Remember that $0.1 = \frac{1}{10}$

Exercise 32C Combinations of operations with fractions

1 Evaluate the following.

a $1\frac{5}{8} + 3\frac{5}{16} + 3\frac{1}{24}$
b $6\frac{7}{16} + 3\frac{3}{7} - 7\frac{7}{20}$

c $\frac{5}{12} \times \frac{6}{11} \div \frac{1}{30}$
d $\frac{3}{5} \times \frac{2}{13} \div \frac{3}{11}$

e $2\frac{9}{24} \times 3\frac{5}{18} \div 1\frac{1}{10}$
f $4\frac{1}{4} \times 3\frac{7}{9} \times 2\frac{5}{13}$

 2 Use your calculator to evaluate the following.

a $6\frac{8}{11} + 2\frac{1}{2} \times 4\frac{1}{6}$
b $(7\frac{2}{7})^2 - 5\frac{1}{4} \div 3\frac{4}{5}$
c $5\frac{3}{4} \div \left(\frac{5}{9} \times 3\frac{3}{4}\right) + 2\frac{1}{3}$

d $\frac{2}{3}$ of $5\frac{2}{7} + 6\frac{1}{2}$

> **Hint** 'of' means multiply.

e $2\frac{2}{5} - 1\frac{1}{4} \div 1\frac{2}{3}$
f $4\frac{2}{3} \times \left(2\frac{5}{8} - \frac{4}{5}\right) + 3\frac{5}{6}$

> **Hint** Remember to apply the correct order of operations.

3 Evaluate the following.

a $\frac{1}{2}\left(\frac{2}{3} + \frac{1}{4}\right)$
b $\frac{2}{5}\left(\frac{3}{4} - \frac{4}{7}\right)$
c $\left(\frac{3}{4} + \frac{2}{5}\right)\left(\frac{2}{3} - \frac{7}{12}\right)$

d $\frac{2}{7}\left(4\frac{1}{2} + 2\frac{3}{4}\right)$
e $3\frac{1}{5}\left(2\frac{2}{3} - 1\frac{3}{4}\right)$
f $\left(2\frac{1}{3} - 1\frac{3}{4}\right)^2$

33 Comparing data using statistics

Example

A supermarket sells oranges in bags of six. The mass of each orange in a selected bag was:

134g 135g 142g 153g 156g 132g

a Find the mean and standard deviation of the mass of the oranges in the bag.

b A second bag was selected and the mass of each orange measured. The mean was 144g and the standard deviation was 8.5g.

Make two valid comparisons between the masses of the oranges in the two bags.

a Mean = $\dfrac{134 + 135 + 142 + 153 + 156 + 132}{6} = \dfrac{852}{6} = 142\,g$

You can calculate standard deviation using one of the following methods.

Method 1: Use $s = \sqrt{\dfrac{\Sigma(x - \bar{x})^2}{n - 1}}$

Method 2: Use $s = \sqrt{\dfrac{\Sigma x^2 - \dfrac{(\Sigma x)^2}{n}}{n - 1}}$

x	$x - \bar{x}$	$(x - \bar{x})^2$
134	−8	64
135	−7	49
142	0	0
153	11	121
156	14	196
132	−10	100
Total	0	$\Sigma(x - \bar{x})^2 = 530$

x	x^2
134	17956
135	18225
142	20164
153	23409
156	24336
132	17424
$\Sigma x = 852$	$\Sigma x^2 = 121514$

$s = \sqrt{\dfrac{\Sigma(x - \bar{x})^2}{n - 1}}$ — This formula is given in the exam.

$= \sqrt{\dfrac{530}{5}} = 10.3\,g$

$s = \sqrt{\dfrac{\Sigma x^2 - \dfrac{(\Sigma x)^2}{n}}{n - 1}}$ — This formula is given in the exam.

$= \sqrt{\dfrac{121514 - \dfrac{(852)^2}{6}}{5}}$

$= \sqrt{\dfrac{530}{5}} = 10.3\,g$

b On average, the oranges were heavier in the second bag than in the first. — Compare the mean using the phrases 'On average' or 'In general'.

The masses varied less in the second bag. — Use phrases such as 'varies more/less', 'more/less spread out' or 'more/less consistent'.

Exercise 33A Mean and standard deviation

1 Find the mean and standard deviation of:

   a 5   8   9   12   15   17   **b** 21   13   16   24   18   21   20

   c 121 136 142 132 139   **d** 2214 2153 2106 2076 2182 2193

2 Find the mean and standard deviation of the following sets of numbers.

a 1 3 3 8 10 (leaving your answer to standard deviation in the form $\frac{\sqrt{a}}{b}$)

b 1 1 2 2 2 3 3 5 5 6 (leaving your answer to standard deviation in the form $\frac{k\sqrt{a}}{b}$)

c 3 4 4 4 10 (leaving your answer to standard deviation in the form $\sqrt{a}$)

d 1 2 2 2 2 3 3 3 4 8 (leaving your answer to standard deviation in the form $\frac{\sqrt{a}}{b}$)

3 The masses, in kilograms, of players in Ayeton High School's football team are:

68 72 74 68 71 78 53 67 72 77 70

a Find the mean and standard deviation of the mass of the players in the football team.

b The mass of players in Beeton Academy's football team has a mean of 68 kg and a standard deviation of 4.24 kg.

Make two valid comparisons between the mass of the football teams of Ayeton High School and Beeton Academy.

4 Jez and his friends participate in a weekly quiz team. Over the past eight weeks their scores were:

62 58 24 47 74 51 60 64

a Find the mean and standard deviation of their scores.

b Rebecca's friends participate in a rival quiz team. Over the past eight weeks their mean scores were 52 and their standard deviation was 9.45.

Make two valid comparisons between the scores of Jez's and Rebecca's teams.

5 In a dance competition, there are six judges, each giving a score out of 10.

a Kathy' scores were: 8 5 6 6 7 4

Find the mean and standard deviation of Kathy's scores.

b Connie's scores were: 9 3 7 10 3 4

Find the mean and standard deviation of Connie's scores.

c Make two valid comparisons between Kathy's and Connie's scores.

6 The number of eggs laid each day by Henrietta the hen were recorded for 1 week. The results were:

1 0 3 2 5 2 1

a Find the mean and standard deviation of the number of eggs Henrietta laid.

b The following week Henrietta's results produced a mean of 1.8 and standard deviation of 1.52.

Make two valid comparisons between the two weeks' results.

7 A group of pupils had a fractions test in October with marks out of 20. Their results were:

8   12   14   6   15   17   11   13

a Find the mean and standard deviation of their test results.

b The pupils attended a weekly revision class, and sat a second fractions test in December, also out of 20. Their results gave a mean of 14 and a standard deviation of 2.14.

Make two valid comparisons between the results in October and December.

8 A travel company sold holidays to 10 couples on a particular day. The mean price of their holidays was £856 and the standard deviation of prices was £32.50.

   a If the travel company added a £5.50 booking fee to every holiday, state the mean and standard deviation of prices after the booking fee was added.

   b Explain how you worked out your answers to part **a**.

Exercise 33B Median and semi-interquartile range

1 For the following sets of numbers find:

  i the median

  ii the lower and upper quartiles, the interquartile range (IQR) and the semi-interquartile range (SIQR).

  a 1   1   4   5   6   7   7   8   10   12

  b 18   14   21   4   17   23   26   15   19

  c 31   14   52   16   71   54   27   65   45   29   50

> **Hint** The lower quartile (Q_1) is the median of the lower half of the data; the upper quartile (Q_3) is the median of the upper half.
>
> The IQR = $Q_3 - Q_1$, and the SIQR = $\dfrac{Q_3 - Q_1}{2}$

2 **a** The ages, in years, of the members of the music department of a school are:

    42   52   53   45   51   55   50   52   49

    Find the median, interquartile range and semi-interquartile range of their ages.

  b The ages, in years, of the members of the English department of the same school are:

    23   41   32   36   29   31   51   33   38   42

    Find the median, interquartile range and semi-interquartile range of their ages.

  c Make two valid comparisons between the ages of the members of the music and English departments.

3 **a** The annual salaries of a group of footballers are:

    £806 000   £2 146 000   £132 000   £649 000   £1 423 000   £425 000   £2 965 000
    £912 000   £1 438 000

    Calculate the median, interquartile range and semi-interquartile range of their salaries.

  b The annual salary of a group of actors had a median of £1 425 000 and a semi-interquartile range of £985 500.

    Make two valid comparisons between the salaries of the group of footballers and the group of actors.

4 Freddy Rocker is a famous singer. The length of the songs on his most recent album were:

  2 min 35 s   3 min 12 s   2 min 48 s   3 min 33 s   2 min 18 s   3 min 45 s
  4 min 3 s   3 min 43 s   4 min 31 s   5 min 11 s   3 min 24 s

  a Find the median and semi-interquartile range of the length of songs.

  b The median of the length of songs on his first album was 3 min 21 s and the semi-interquartile range was 49.2 s.

    Make two valid comparisons between the length of songs on Freddy's first album and on his most recent album.

5 Holly researched travel insurance for a 5-day trip to Paris and found the following quotes from different insurance companies.

£5.25   £6.78   £10.15   £8.40   £5.00   £7.86
£9.64   £8.24   £11.55   £7.50   £6.42   £7.00

a Find the median and semi-interquartile range of these quotes.

b Holly's friend Jim researched travel insurance for a 5-day trip to New York. His quotes had a median of £9.42 and a semi-interquartile range of £4.96.

Make two valid comparisons between the quotes for travel insurance for trips to Paris and New York.

6 A college rugby team counted the number of press-ups they could complete in one minute. The results were:

65   70   62   58   72   80   64   62   71   85   90   74   67   71   72

a Find the median and semi-interquartile range of their press-ups.

b The college football team also counted the number of press-ups they could complete in one minute. They calculated a median of 66 and a semi-interquartile range of 8.5.

Make two valid comparisons between the press-ups completed by the football and rugby teams.

7 A group of students were asked how long it took them to complete a maths homework task. The results were:

45 min   32 min   35 min   38 min   48 min   1 hour 2 min   50 min   36 min

a Find the median and semi-interquartile range of the length of time it took to complete the maths homework task.

b The same students were asked how long it took them to complete a history homework task. The median of their results was 45 min and the semi-interquartile range was 6 min.

Make two valid comparisons between the lengths of time it took for the students to complete their maths and history homework tasks.

8 Janet sat exams in eight different subjects. Her results were:

English 68%   Maths 72%   French 64%   Chemistry 70%

History 74%   Music 42%   Art 62%   Computing 76%

a Find the mean and standard deviation of her results.

b Find the median and semi-interquartile range of her results.

c Which measure of average and spread should be used in this question? Give a reason for your answer.

34 Forming a linear model from a given set of data

Example

The table shows the heights and masses of 12 students in a class.

Student	Mass, A (kg)	Height, H (cm)
Arianna	41	143
Bea	48	145
Caroline	47.5	147
Dhiaan	52	148
Emma	49.5	149
Fiona	55	149
Gill	55	153
Hanna	55.5	155
Imogen	61	157
Jasmine	65.5	160
Keira	60	163
Laura	68	166

a Plot the data on a scatter graph.

b Draw the best-fitting line.

c Find the equation of the best-fitting line in terms of A and H.

d A new girl with a mass of 45 kg joined the class. Use the equation of the best-fitting line to estimate her height.

e Chloe was absent from the class when they were measured, but is 152 cm tall. Use the equation of the best-fitting line to estimate her mass.

a and b

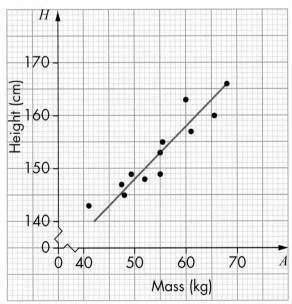

c

Gradient, *m*

Using coordinates (55, 153) and (68, 166):

$$m = \frac{166 - 153}{68 - 55} = \frac{13}{13} = 1$$

> **Hint** To find the gradient (*m*), use two points that lie on the line.

There are two methods you can use to work out the equation of the best-fitting line.

Method 1: Use *y = mx + c*

$y = (1)x + c$ — Substitute the value for *m* into the equation of a straight line *y = mx + c*.

$153 = 55 + c$

$c = 98$ — Use the *x*- and *y*-values of one of the coordinates (i.e. 55, 153) and substitute these into the equation to find *c*.

$y = x + 98$

$H = A + 98$ — Substitute *c* into the equation.

— Change the variables to *H* and *A* in place of *y* and *x*.

Method 2: Use *y − b = m(x − a)*

$y − 153 = 1(x − 55)$ — Substitute gradient *m* = 1 and the *x*- and *y*-values of one of the coordinates (i.e. 55, 153) into the equation to find *c*.

$y − 153 = x − 55$

$y = x + 98$

$H = A + 98$ — Change the variables to *H* and *A* in place of *y* and *x*.

d $H = 45 + 98 = 143\,\text{cm}$

e $152 = A + 98$

$A = 54\,\text{kg}$

Exercise 34A Drawing and using a best-fitting line from given data

 1 The table shows the marks for 10 students in their mathematics and music exams.

Student	Alex	Ben	Chris	Dom	Ellie	Fiona	Gordon	Hannah	Isabel	Jemma
Maths	52	42	65	60	77	83	78	70	29	53
Music	50	52	60	59	61	74	64	68	26	45

a Plot the data on a scatter diagram. Use the horizontal axis for the scores of the maths exam and mark it in 20s from 0 to 100. Use the vertical axis for the scores of the music exam and mark it in 20s from 0 to 100.

b Draw a best-fitting line.

c Find the equation of the best-fitting line in terms of *y* and *x*.

d Kris was absent for the music exam but scored 45 in the maths exam.

Use the equation of the best-fitting line to estimate his score in the music exam.

e Lexie was absent for the maths exam but scored 78 in music.

Use the equation to estimate her mark in the maths exam.

2 The table shows the time taken and distance travelled by a delivery van for 10 deliveries in one day.

Distance (km)	8	41	26	33	24	36	20	29	44	27
Time (min)	21	119	77	91	63	105	56	77	112	70

a Draw a scatter diagram with time on the horizontal axis.

b Draw a best-fitting line.

c Find the equation of the best-fitting line in terms of D and T.

d A delivery takes 45 minutes. **Use your equation** in part **c** to estimate the distance travelled.

e **Use your equation** to estimate the length of time taken for a distance of 30 km.

3 Harry records the time taken, T, in hours, and the distance travelled, D, in miles, for several different journeys.

Point A represents a journey of 78 miles taking 2 hours.

Point B represents a journey of 195 miles taking 5 hours.

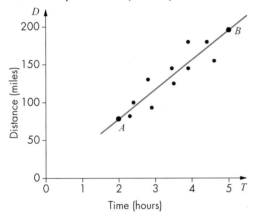

a Find the equation of the best-fitting line.

b **Use the equation** to estimate the distance travelled for a journey lasting 3 hours.

c **Use the equation** to estimate the time taken for a journey of 175 miles.

4 Twelve students took part in a maths challenge. There were two parts: a mental test and a problem-solving test. The scatter graph shows the relation between the two parts.

Point C represents a student who scored 12 in the mental test and 17 in the problem-solving test.

Point D represents a student who scored 40 in the mental test and 38 in the problem-solving test.

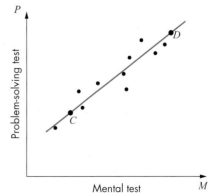

a Find the equation of the best-fitting line in terms of P and M.

b **Use the equation** to estimate the score in the problem-solving test of a student who scored 32 in their mental test.

c **Use the equation** to estimate the score in the mental test of a student who scored 23 in their problem-solving test.

5 Pupils in a class measured their height and the length of their feet. The scatter graph shows the relation between the length of their feet, F (in cm), and their height, H (in cm).

Point E represents a pupil with a foot length of 25 cm and a height of 146 cm.

Point F represents a pupil with a foot length of 33 cm and a height of 170 cm.

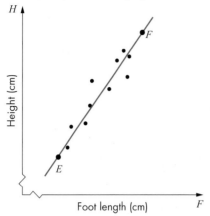

a Find the equation of the best-fitting line in terms of H and F.

b **Use the equation** to estimate the height of a pupil with feet of length 28 cm.

c **Use the equation** to estimate the length of feet of a pupil of height 164 cm.

6 The week prior to the SQA exams, a group of pupils were asked how long they spent studying and how long they spent on leisure activities. The scatter graph shows the relation between their revision time (R) and their leisure time (L).

Point G represents a pupil who spent 9 hours revising and 33 hours on leisure activities.

Point H represents a pupil who spent 41 hours revising and 17 hours on leisure activities.

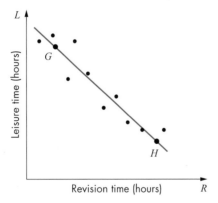

a Find the equation of the best-fitting line in terms of R and L.

b **Use the equation** to estimate the length of time spent on leisure activities by a pupil who revised for 22 hours.

c **Use the equation** to estimate the length of time spent on studying by a pupil who spent 36 hours on leisure activities.

7 A shop recorded their glove sales and the maximum outside temperature every day during November. The scatter graph shows the relation between the temperature, T (°C), and the number of pairs of gloves sold, N.

J represents a day when it was 4 °C outside and they sold 22 pairs of gloves.

K represents a day when it was 10 °C outside and they sold 4 pairs of gloves.

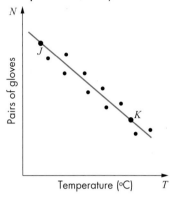

a Find the equation of the best-fitting line in terms of N and T.

b **Use the equation** to estimate the number of pairs of gloves sold if the temperature is 6 °C.